Theoretical Principles of Heart Cycle Phase Analysis

M. Rudenko, V. Zernov, D. Makedonsky,
K. Mamberger, O. Voronova, S. Kolmakov,
S. Rudenko, A. Volkov, N. Volkova,
O. Volossatykh, S. Khlestunov,
Y. Prikhozhan

Theoretical Principles

of Heart Cycle Phase Analysis

Part I
Theory of Hemodynamics and Metrological Support
Principles when Measuring
Cardiovascular System
Phase Parameters

Part II
Atlas
of Functional Diagnostics Based
on Heart Cycle Phase Analysis
(ECG + RHEO)

FOUQUÉ PUBLISHERS NEW YORK

Copyright ©2011 by Fouqué Publishers New York
Originally published as *Mäuslegeschichten, 2006*
by August von Goethe Literaturverlag

First American Edition
Printed on acid-free paper

Library of Congress Cataloging-in-Publication Data
[Theoretical Principles of Heart Cycle Phase Analysis. English]
1st American ed.

ISBN 978-0-578-09470-0

The authors would like to acknowledge the individual contribution to the cause of publication of this book made by:

Mrs. Larissa D. Rudenko,
Senior Project Economist (Taganrog, Russia);

Mrs. Galina M. Straussova,
Chief Accountant, Russian New University (Moscow, Russia);

Mr. Vladimir B. Alexeyev,
Cosmonaut (Moscow, Russia);

Mr. Boris J. Leonov,
General Director, The Russian Research & Development Institu-
tion VNIIMT
(Moscow, Russia);

Mr. Vyacheslav V. Akhromushkin,
Exhibition Committee Manager, VVTs (Moscow, Russia);

Mr. Mikhail J. Reva,
Senior Lecturer, the St. Petersburg State Technical University (St.
Petersburg, Russia);

Mrs. Marina Kutuzova & Mr. Sergey Kutuzov,
Co-worker of the R & D Venture Enterprise NT OOO «Cardiocode» (Taganrog, Russia);

Mrs. Hermine Tenk,
Dr. med., Professor (Baden, Austria);

Mrs. Jaanna Koponen-Kolmakow,
Director, Oy Cardiocode – Finland Ltd. (Kuopio, Finland);

Mr. Hannu Janhunen,
Managing Director, Technopark «Teknia» (Kuopio, Finland);

Mr. Ilkka Vartiainen,
Development Manager, Technopark «Teknia» (Kuopio, Finland);

Mrs. Tamara Weber & Mr. Konstantin Weber,
Directors, «Cardiocode Ltd. & Co. KG» (Simmern, Germany).

We admire Nature's Engineering
in the most perfect cardiovascular system
that is left to fate of humanity
for endless study for ages.
In c ommem orati on
of our esteemed Colleague
Mr. Gustav M. Poyedintsev
who discovered
the Laws of Hemodynamics.

A distinguishing feature of the book is the description of theory for cardiac cycle phase analysis. The analysis is based on a new fundamental scientific discovery which explains the blood flow in major blood vessels. Formerly, the theory of heart and vessels work was based on laminar blood flow pattern. That theory did not account for some important pfenomena and contradicted the real facts. The erroneous theories were used in practical medicine that kept from establishment of metrological basis for diagnostic technique with ECG tracing being in the first place. All known noninvasive methods of heart and vessels diagnostics belong to the group of indicating means rather then to measuring instruments.

The theory of high fluidity condition of blood enabled the medical science to approach the metrological aspect of diagnostics based on ECG and rheogram methods.

The phase analysis method described in the present book is the first step to the differential diagnosis of the heart and vessels work. The method offers the early and more accurate determination of cause of heart and vessels pathological condition.

Up to the present time, all methods were based upon the integral factors that allowed the recognition of the diseased condition only post factum. The theory set forth in this book enables the physicians to predict the pathological process that has been absolutely impracticable earlier.

We hope that our work will contribute to further development of cardiology theory. As experience has shown, the application of the described theory enables an average practicing physician to work successfully against the various manifestations of cardiovascular system diseases.

Authors

Introduction

Every second, the heart of a person does the work that shall be maintained by the human organism in cycles throughout the entire life. Cardiac arrest emergency causes instantaneous lethal outcome. Even the localized cardiac pathology hinders greatly normal functioning of human organism and quite often incapacitates the person. It is impossible to overestimate the importance of normal functioning of cardiovascular system because heart and vessels are main blood transporting systems that provide necessary nutrients to every single cell of the organism.

In spite of significant scientific-and-tech-nological advance, there has still been no medical diagnostic equipment that would help to make reliable diagnosis of cardiovascular system functions and hemody-namics parameters. In general, measurement problem really exists in the medical instrument making industry, and in this respect we have a crisis situation. This is caused by a lot of factors. One of these factors, and the most crucial one, is that it is difficult to represent biophysical processes in medical science. Actually, there has still been no one model that would adequately reflect basic properties of any physiological organ, which could be assumed as reference object when carrying out the measurements. Nowadays, only thermometry and blood pressure measuring procedures meet to some extent the requirements for physiological parameters measurement. Therefore, to solve the problem of measurement efficiency, a new physiology knowledge paradigm is required. This paradigm shall be based on theoretical models, which are abstracted to the level of basic functions of each system. Thus, cardiology requires the availability of a model that will explain energy processes of hemody-namics. Such hypothetic models shall gain the status of axioms, without which it is impossible to develop theoretical concepts. The time period for solving this problem is not determined yet.

As a consequence, there has been no uniform understanding of ischemic heart dis-ease[1] [1] criteria so far. Myocardial infarction can be diagnosed only in four hours after its occurrence[2] [2], while in practice the effective medical assistance can be rendered only within first twenty minutes after beginning of pathological process.

Nowadays, it is impossible to predict the state of cardiovascular system. All hardware enable recording of only those events that have already taken place, and it is practically impossible to assess the pathogenic mechanism and predict the course of some disease.

In social implication, this situation does not make possible to accomplish a credible health survey and effective preventive medical examination of population. Let us consider insurance medicine of European Community countries. The expenditures that one way or another are connected with solving the problems related to cardiovascular system treatment constitute up to 30 percents of total medical treatment costs.

There are dozens of special cardiological programs in the world, these programs are financed by government and aimed at solving the global problems of cardiovascular system disease prevention. However, results which have been achieved are still very poor.

A lot of knowledge of heart and heart and vessels interaction is obtained from surgical practice. Although efforts to create artificial heart did not reach the goal, they brought to light a paradox of myocardium cells self-regeneration which can occur under certain conditions.

In general physiology, study of physical activity hemodynamics has shown that the range of healthy heart performance essentially overlaps the range of performance of heart with pathology. This obscures the borderline between norm and pathology. From our point of view,

1   A New Definition of Myocardial Infarction (a joint document of the Unified Committee of the European Cardiology Society and the American Cardiologists Collegium with respect to revision of the existing definition of myocardial infarction)// Journal of the American Cardiologists Collegium (JACC), Vol.36 No.3, 2000, September, P.959-969.
2   Beryozov T.T. Application of Ferments in Medicine // Soros' Education Journal, Biology. 1996 No. 3, P.23-27.

this is one of the main factors that caused conceptual difficulties in development of cardiologic theory.

In spite of total effort to enable measurement of hemodynamics parameters, the scientific world only came nearer to general solution of individual problems. In particular, the problem of measuring the phase characteristics of cardiac cycle is recognized to be a priority issue. In opinion of leading experts of the world, phase characteristics represent all features of hemodynamics processes in normal and pathological cases[3] [3].

The present paper covers results of research in the field of cardiovascular system parameters measurement that has been carried out within period of a quarter of a century. Within the period of the work on this scientific research paper, two scientific discoveries were made. One of them has been recognized by the public[4] [4] , a n d the second one is under expert evaluation and has been registered as application for a discovery. It is also necessary to mention the fact that the problem of hemodynam-ics characteristics measurement has been solved thanks to one more unique discovery of blood superfluidity in major vessels. This laid the foundation of present paper. Moreover, thanks to this very discovery, the most complex hemodynamics processes have been simulated, and heart phase transition parameters measurement has been enabled in terms of metrology.

---

3   Caro C, Padley T., Shroter R., Sid W. Circulation Mechanics, M.: "Mir", 1981. P.624

4   Scientific Discovery No.290 Law of Propagation of Arteric Pressure Waves in Blood Vessels in Areas of Local Increase in their Impedance / Alexeyev V.B., Zernov V.A., Matsyuk S.A., Rudenko M.Yu.. Application dd.05.07.2005, A-358; published 31.08.2005 [electronic resource:] Access: http://www.raen.ru/discovery/288.shtml7in

Part I

Theory of hemodynamics
and principles of cardiovascularphase
parameter measuring

1. State of electrocardiosignals research methodology

1.1. Conceptual issues of medicobiologic signals research methodology

The issue of uniformity of measuring tools in medical science is practically not closed so far. The reason for this is the infeasibility to simulate physically the reference models of organs and organs functions which could be used as reference measuring tools. This situation hinders development of medical instrument making industry to a great extent. Even those diagnostic units which are in series production do not meet the measuring instruments uniformity requirements. It goes without saying, that the lack of reliability and low confidence level in medicobiologic signals recording lead to the liberal interpretation by authors of the same data and findings. Mere administrative introduction of standards makes implementation of research findings even more complicated, since it makes unshakable those concepts which should be revised due to the engineering and technological advance.

The question arises whether it is at all pos-s ible t o atta in the u n i for m ity of mea sur i ng tools within the medical instrument making industry. To our opinion, it is possible if we follow the most difficult, but the only way available for now. Stages of this way are shown in block diagram (Fig. 1).

The matter can be tackled through theoretical development of the hypothetic models of biophysical processes under study. The development of any model needs to be started with the development of a theorem. As a rule, the development process itself takes a lot of time.

The main objective of theorem formulation is to isolate characteristic features of the phenomenon under study from results of experiment. This process consists in abstracting (generalizing) to the level of properties peculiar to constituent parts of the phenomenon. The isolated properties must exactly suit the context of phenomenon general description. Through this differentiation we obtain a theoretical construction, or theoretical model (to be more precise), which represents basic properties of the phenomenon under study. However, such theoretical model requires a larger number of practical experiments.

Afterwards, the process can be developed in two ways. The first way is to apply theoretical proof; the second way is to apply practical proof in order to check characteristic features of the selected axiom. It is desirable that theoretical proof should contain mathematical analysis. The proof implies that the theorem formulated in this way shall be checked for being consistent with the existing ones. It must fall into a series of theorems which build up the theoretical foundation for the given field of knowledge. This process requires public discussions and stimulates the development engineers and designers to publish their papers in periodicals, and in some cases to issue monographs.

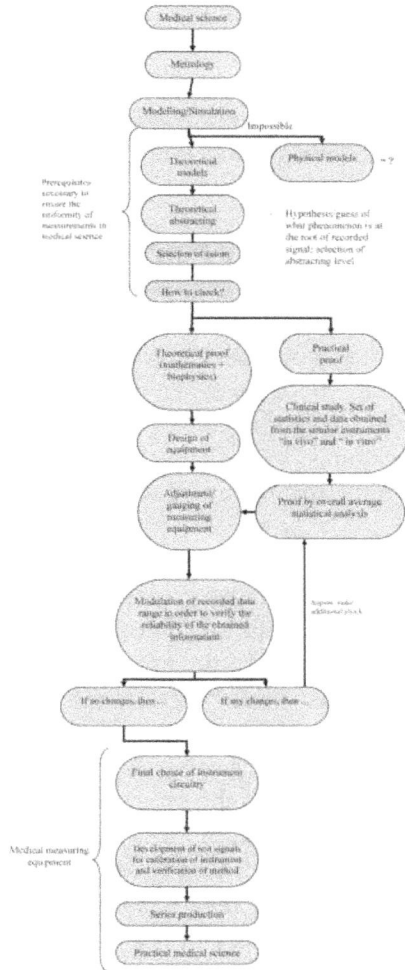

Fig. 1. Model for solving the problem of measuring tools uniformity in medical instrument making industry

Practical proof is as follows. It is necessary to obtain real results of measurements carried out with use of existing instruments. It is of no importance whether the instruments are metrologically-supported or not. What really matters is that they must be practiced in medical procedures. Simultaneously, it is necessary to accumulate statistical data by practicing the suggested method. When collecting the statistical data, particular emphasis shall be placed upon the norm-pathology borderline cases. Hereafter, in case the results do not conflict with the real values, it is necessary to refine theoretically the shape of the recorded signal, for various aspects of normal and pathological cases.

The numerical values obtained through measurements are processed with use of statistical methods, thus we reveal the domain where the probability of existence of normal values corresponding to normal state is the highest.

But these procedures do not complete the process. The next step is to establish the range within which the recorded signals are most reliable.

If there are any doubts in reliability of signals (which is usually the case), the parameters of instrument electronic circuitry must be changed and procedure must be repeated. When positive results are obtained, we can proceed to the final schematic design of circuit of the instrument and to development of test signals.

As a rule, in the course of proving, the reasons are shown for the refinement of the wording of the previously formulated theorem and for selection of some axiom. The cause for this is that both theoretical concepts and experimental data are used for proving. But experimental data shall be obtained through use of more than one instrument manufactured as a development batch.

It is very important to record experimental environment and conditions which is essential in processing the data selected for the proof. Everything which is directly or indirectly associated with data acquisition must be described in detail. It may be required for further certification.

As the final result, test signals are to be used as reference signals for verification or calibration of equipment. Actually, this approach can facilitate the equipment certification. Biophysical process theorem can be a substitute for the reference metrologi-cal model. This specifies the significance and necessity of theorem formulation.

The methods to be used for diagnostics of some specific symptom follow from the theorem. While the theorem comprises proof and axiomatic formulation, the methods as a rule are based only on its conceptual statements.

It is not of less importance that more precise measurements often offer a revisions for the existing theories of biophysical processes, these revisions can prove themselves to be essential. In fact, it leads to development of some new theory and directly affects the paradigm of knowledge in the field of research. As consequence, more time can be required for the instrument certification.

The above-mentioned issues are not the only ones related to methodology of medi-cobiologic signals research. Many other minor factors should be taken into consideration. So, the investigation of various natural phenomena requires employment of dedicated measuring instruments. As a rule, they are very expensive and not always affordable for development engineers and designers, moreover, employment of these specific instruments increases the cost of development.

Preliminary clinical testing also requires the time to accumulate the statistical data on normal and pathological conditions of patients. The patients shall be provisionally diagnosed by the established methods. All data received in the course of experiments shall be classified. One can be sure in the reliability of diagnostic performance of a new instrument only after the comprehensive analysis of data obtained through practical work.

1.2. Current methods for cardiovascular system investigation.

Just over one hundred years ago the instruments came into use for investigation of cardiovascular system condition. It is a very short space of time for the science. The law formulated by Faradey laid the foundation for engineering and design of various electrically-driven mechanisms, it also motivated the employment of technological and scientific achievements in diagnostics and medical care. A heart is a unique natural mechanism, secrete of heart work stimulated research activities of scientists at all times from the ancient world up to now. But only discovery of laws related to motion of charged particles in electrical circuits provided the real possibilities for initiation of more detailed investigation of principles of cardiac activity. The first attempt to record an electrocardiosignal (which is actually a prototype of nowadays ECG tracing) was undertaken by W. Einthoven in 1912 in Cambridge[5] [5]. From then on, the methods of ECG tracing progressed in fast pace. Simultaneously, new methods of cardiovascular system investigation were developed. First of all, phonocardiography has been put into practice. This method was based on recording the cardiac murmur existing during heart work. The phonocardiography arouse from auscultation of body practiced in the earliest times. Also, invented was a first arterial blood pressure meter operation of which was based on indirect method of measurement[6] [5] .

Invention of conductive materials allowed recording of body bio-potential and body mechanical parameters. So, the first sensors for ECG tracing invented by Einthoven (which were actually the buck-ets filled with water and connected to make a galvanic cell when hands and legs were dipped into water) were replaced in 1917 with

---

5   Medical Electronic Equipment for Public Health Service, Transl. from English into Russian, M.Arditti, by F.Waybell et al., Translation edited R.I.Utyamysheva, M., „Radio I Svyaz", 1981.
6   In the same reference.-P. 135-144.

metal electrode plates[7] [5]. Similar electrodes are in use now in some health centers. They were made from metal which was covered with thin layer of silver chloride. This coating enabled to make the contact potential difference on metal - skin interface more stable.

In postwar period from 1945 on, radio electronics was in good progress, the various specialized diagnostic tools came into use. In design of instruments, the electronic/mechanical element base was employed. First of all, the arterial blood pressure meters should be mentioned[8] [6] . T he proposed indirect method of measurement is based on the arterial pressure monitoring by the scale of mechanical manometer. This method is widely used until now.

Beginning from 1960, quantity of electronic components in medical instruments increases. The industry launches a serial production of complicated instruments to which unquestionably electrocardiographs belong. The instruments incorporate electronic filters for filtering the recorded signals. Special-purpose mechanical recorders were used to record the ECG trace on paper.

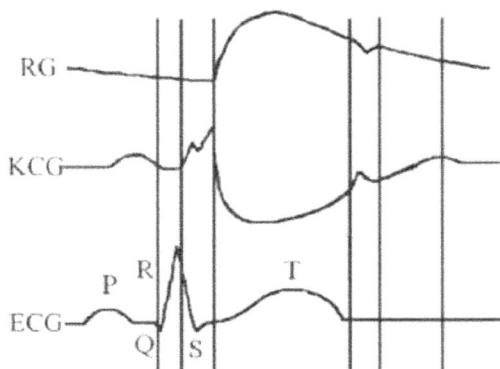

Fig. 2. Polycardiographical method: synchronous recording of rheogram (RG), kinetocardiogram (KCG) and ECG trace

---

7   In the same reference. - P. 109
8   Shelagurov A.A. Propaedeutics of Internal Diseases. M.,"Medicina", 1975.

However, the value of measured arterial pressure and ECG trace were not enough for reliable diagnostics. At the same time, the multi-channel ECG recording was invented, policardiography methods for cardiovascular system examination came into use[9] [6]. They were based on synchronous registration of electrocardiosignals and of a number of physical characteristics of heart: a rheogram[10] [6] was used to show the vessels filling with blood and arterial pressure fluctuation; kinetocardiogram was used to show mechanical fluctuations[11] [7], or ballistocardiogram[12] [6] was made for the same purpose which was more complicated procedure since a special sensor was used (Fig. 2).

Possibility of recording various biological signals generated a vital issue of authenticity of diagnostic criteria contained in the signals. Even the first practical outcomes evidenced that the accuracy of biological signals recording must be much higher. Any artifacts, existing during signal recording procedure, prevented from obtaining the identical signals during repeated procedure. The required level of accuracy was not achieved even when the doctor thoroughly observed the complicated procedure for instrument application.

For example, to analyze the phase work of the heart, a phase boundaries must be recorded, in doing so the phase must be chosen duration of which do—es not exceed 0.05 s. Polycardiography method and ECG multi-channel recording method have been developed for this purpose. Jet pens used for recording have the thickness commensurable with accuracy required for comparative analysis. Multi-channel read-out of data increases the measurement error even more due to asynchronous processes occurring in various sensors which differ in their principles of operation and are used to register different electric and mechanical characteristics of heart work.

---

9   In the same reference. P. 219.
10   In the same reference. P. 179-185.
11   Andreyev L.B., Andreyeva N.B. Cinetocardiography. Rostov-on-Don, Published by the Rostov State University, 1971, 308 pages.
12   Shelagurov A.A. Propaedeutics of Internal Diseases. M.,"Medicina", 1975C. 213-215.

The problem was not resolved even when computers came into use in medical science. The same sensors which have been used earlier, are still in use. They are sources of errors. Requirements for accuracy of medicobiologic signals measurement have been always ahead of technological development. This fact set forth the issue of metrological examination of medical instruments. Clinical testing came to the topside of instrument certification procedure, since the conclusions about authenticity and reliability of measurements results are mostly based on statistic data, not on calibration testing. Synchronous recording of a number of signals of different nature can only serve as an indicator of body condition, but it can not be used as a measuring system.

So, this was the situation in which the methodology for cardiovascular system investigation has been developed.

The medical instrument making industry advances, the vacuum cells were super-ceded with solid-state semiconductor elements, but the issue of safe employment of electrically-driven instruments has been still in the scene[13] [8]. Galvanic decouplers made on transformers and ground loops were interfering sources and essentially decreased the quality of recorded signals.

Invention of dopplercardiometry (echocar-diography) in the eighties was a significant advanced development in the field of cardiovascular system diagnostics. The dop-plercardiometry methods were based on cardiac ultrasonic echoscopy (ultrasonic scanning)[14] [9]. The instruments employed in the method are more complicated than those used earlier, but their practical application enabled to expand the diagnostic ranging and quality. In addition to that, hemodynamics values can now be measured using indirect method, e.g. volume of blood per one heart beat and per minute in ascending aorta. Investigation of morphological structure (characteristics) of

13   Popetchitelyev Ye.P, Korenevsky N.A. Electrophysiological & Photometric Medical Equipment. M., "Vysshaya Shkola", 2002, P.208-212

14   Strutynsky A.V. Echocardiogram: Analysis & Interpretation. M., "Medpressinform", 2001, 208 pages

body organs and heart tissue now was possible and is a considerable development thanks to invention of ultrasonic scanning (Fig. 3).

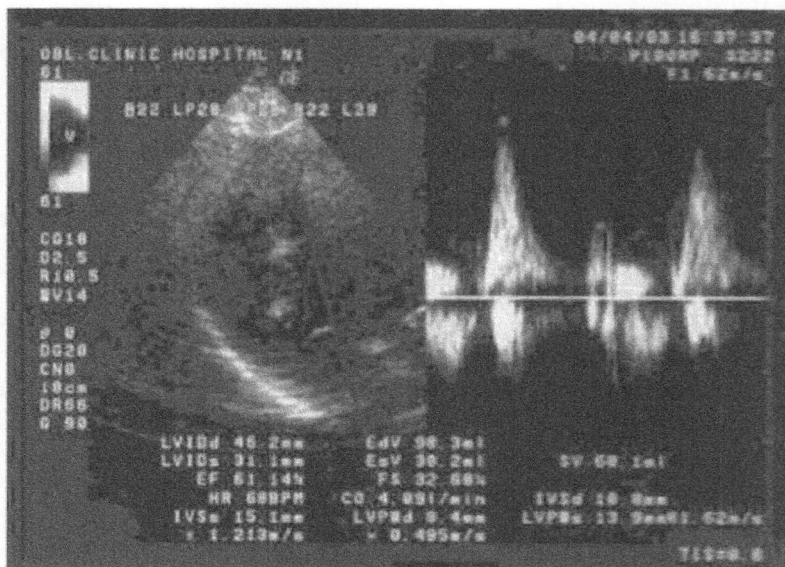

Fig. 3. Visual image obtained through ultrasonic scanning of heart and shown on instrument display

Employment of computers brought in a new era and gave an impulse to instrumentation technology. Now, large blocks of information can be processed with results being displayed simultaneously in various formats, both in digital and graphical.

However, the employment of computers did not allow to say that specialists are nearer to being able to solve the main problem of instrumentation technology that is the metrological guarantee. Both methodology, which is mainly based on data read-out, and data processing, which is actually the signals conditioning and is aimed at obtaining the possibly larger dynamic range of signal-to-noise ratio, remain the same.

Internet and mobile wireless communication facilitate the access to the information source and transmission of information recorded on-line when the patient is being examined. But in spite of intensive technological development, the measurement tools uniformity has not yet been ensured.

Actual diagnostic techniques existing for the moment were mated with computer hardware. Nowadays, diagnostic software/hardware systems can be built based on specialized processors. This way enables to avoid the extra expenditures for standard equipment which in any case can not be used in full scope for body diagnostics purposes but is widely used in other fields of human activities.

This approach predetermined the introduction of small-size high-efficiency software/hardware devices. For their effective employment only the software should be updated without considerable spending on new hardware.

Leading edge technological processes afford the serial production of high performance disposable electrodes. The effect of skin-to-electrode contact potential difference on signal to be recorded was minimized.

The instruments for indirect measurement of arterial pressure are upgraded and become more profound at the same time as the electrophysiological methods of cardiovascular system diagnostics developed.

At present, a great number of instruments of individual use go into quantity production. They are designed to monitor the arterial pressure and heart rate. But all these instruments register the arterial pressure using criteria which are shaped based on overall average analysis. The instrument incorporates an amplitude comparator, which comes into action in response to changing of amplitude of arterial pressure variation under a compression band, thus fixing systolic and di-astolic pressure[15] [10]. This method leaves the stroke volume of blood out of regard, while this parameter should be taken into account since it affects the signal shape. So, in this case we can

---

15   Digital Blood Pressure Measuring Instrument. Catalog: "Sensors produced by MOTOROLA". M., "Dodeca", 2000 P.35-37

speak only about conventional values of measured pressure, and not about the real ones.

The indirect method of measurement gives an error which depends on speed of decompression, it also has an error induced by using the criteria which are based on statistic data. This is why the designers have to introduce restrictive conditions for instruments employment, such as patient position during pressure measurement and portion of arm where the occlusive band is to be fixed.

It is essential to mention the only one instrument existing for the moment which measures the arterial pressure using a crucially new method. The method is based on analysis of arterial pressure waves interference in area of occlusion. The instrument was designed in 1989 based on serial production instruments. The development of theory of biophysical processes in occlu-sive blood flow[16] [11] preceded the design and development of the instrument. At present, the instrument undergoes some modifications and soon will be ready for putting into series production. The restrictions essential for other known methods do not apply to this instrument.

It is important to note that, development of theory of biophysical process in occlu-sive blood flow facilitated the model development, which in its turn ensured the metrological guarantee and high accuracy of measurement. Up to now, none of the instruments for cardiovascular system diagnostics had been designed based on the theoretical model of biophysical processes which would enable to simulate the registered signal thereby providing for accuracy and uniformity of full-scale measurement process.

The lack of models representing the body organs and their functions is the cause of the fact that all the electrocardiosignal-fixing instruments allow only to record the signal but not to measure it. The proper measuring process must be based on comparison of the value to be measured with the reference value. There is no measurements without reference value or measurement standard. We can build up

16   Rudenko M.Yu., Alexeyev V.B., Matsyuk S.A. Biophysical Phenomena in Blood Circulation in Indirect Measuring Arterial Pressures & Evaluation of the Relevant Instrumentation // "Medtechnica". 1986 No.5   P.26-35

a model of biological signal to be taken for reference signal only if we manage to abstract the object. To be more specific, we have to build a model which represents the main properties of the object which are manifested in the phenomenon subjected to investigation. For the time being, this issue remains open[17] [12].

The present situation demands fresh approaches to metrological issues. In this connection it is reasonable to consider the existing paradigm of scientific knowledge in the field of medical science.

17   Popetchitelev Ye.P, Korenevsky N.A. Electrophysiological & Photometric Medical Equipment. M., "Vysshaya Shkola", 2002, P.75-76

1.3. Paradigm of hemodynamic parameters measurement in cardiology

Any paradigm comprises a system of theoretical, methodological and axiological mental sets that are considered to be a pattern for scientific problems solving and that are shared by all members of the scientific community.

In the cardiology, the paradigm has its historical roots. The above-mentioned methods for cardiovascular system investigation can be arbitrary divided into three following categories:

a) The first category, which is a common one, comprises methods that only allow to indicate a parameter. Using these methods, the specialist can take only common decisions like "yes" or "no".

b) The second category includes methods that allow to record cardiosignals. Using these methods, the doctor can diagnose any disease post factum (after the disease has already occurred). But the development of cardiovascular system state can be scarcely predicted.

c) The third category comprises measuring methods. These methods allow to measure parameters. Using these methods the specialist can diagnose and predict the development of the cardiovascular system state. The distinctive feature of this category is that it is based on the results of the basic research.

Let us consider in more detail the grounds that stimulated the development of the measurement paradigm in the cardiology.

Schematically, the aspect of the diagnostic units engineering can be represented in general outline as shown in Fig. 4. The overlap area that constitutes the sector shared by three circumferences represents the feasibility of indicating, recording or measuring medical instruments design. It is impossible to design the cardiologi-cal measuring systems until the special theory for simulation of hemodynamics biophysical processes is developed. This special theory depends on the level of systematic comprehension of human organs and functions interaction.

28

The descent from general to special theory is the main condition under which the processes under study can be simulated.

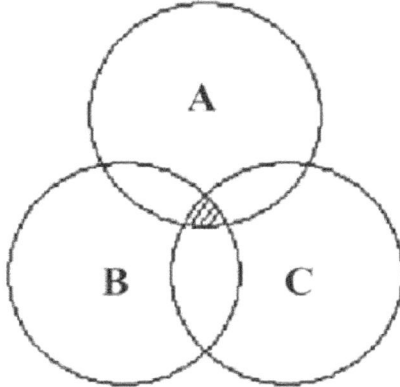

Fig. 4. A defines the general system knowledge,
the paradigm (global information field),
B defines the engineer's knowledge,
C defines the doctor's knowledge

Having received any task assigned by the doctor, the engineer, in his turn, can solve this task only using the existing technical capabilities. At that, he has to refer to the global knowledge in general, to the paradigm that can contain the solution of the scientific problem. But the paradigm does not always contain the model that describes the processes under study. Therefore, the work of the representatives of two fields of knowledge (i. e. the doctor and the engineer) shall include such an aspect as the check of the obtained results for compliance with the standard ones. Finally, the following components shall be available for solving the measurement problem:
1. The theoretical model of the biophysical process under study. This model shall adequately represent the basic features of the biophysical process and shall not conflict with the general laws of the substance development.
2. The theory of the biophysical processes that form the information criteria in the recorded signal.

3. The theoretical and practical proof of the proposed theory.

Required parameters can not be measured if the above-mentioned components are not available. The designers can only provide the means for signals recording, further evaluation and diagnostics shall be made by the doctor. At that, the doctor's knowledge and experience will be of crucial importance when making diagnosis.

The methods for cardiovascular system investigation, which are described in the preceding section, show that the development of the theory in the cardiology hardly comes up to the level of signals recording. Only ultrasonic instruments can be qualified as measuring systems.

The model, that describes biophysical processes and that is taken as a reference, is a pure mathematical model.

In particular, the stroke volume in the ascending aorta is measured using Tei-cholz method[18] [9].

$$V = \frac{7 \cdot 0 \cdot D^3}{(2.4 + D)} \tag{1}$$

where: V is the volume of the left ventricle of the heart; D is A-P dimension of the left ventricle of the heart during the ventricular systolic time interval or diastolic time interval.

The stroke volume is defined as the difference between the end systolic volume and the end diastolic volume[19] [9].

Besides, there is a more complicated equation, so-called Simpson equation, in the ultrasonic scanning that allows to calculate the stroke volume in the ascending aorta[20] [9].

$$V = \frac{\pi}{4} \cdot \Sigma (a_i + b_i) \cdot \frac{L}{20} \tag{2}$$

---

18   Strutynsky A.V. Echocardiogram: Analysis & Interpretation. M., "Medpressinform", 2001,   P. 76.
19   In the same reference.
20   In the same reference.

In practice, the first method is used more often than the second one, because it requires less time for the patient examination.

However, out of many hemodynamic parameters, the ultrasonic instruments measure only the stroke volume. As for the rest, the ultrasonic instruments measure the linear dimensions of the organs. Besides, they determine the morphological state of the tissues based on the tissue density that affects the echo signal absorbtion.

While considering the measurement aspect in cardiology, it is necessary to note again that development of the theory of the biophysical processes, which form information criteria in the recorded signal, depends on the measurement accuracy of the used instruments. It is impossible to study the required biophysical processes and to improve the theory until the required accuracy of signal recording is ensured. For this particular matter, the general level of technological development of society plays a crucial role. Introduction of ultrasonic medical instruments and radically new methods of body examination can serve as an example.

But let us note that the ultrasonic method does not furnish the solution of problems related to hemodynamic parameters measurement. The polycardiographical method that was presented earlier is aimed mostly at evaluating the phase characteristics of the cardiac activity. This method is considered to be the most promising one in terms of the cardiovascular system diagnostics[21] [13].

The aspect of cardiosignal shape recording accuracy, which also concerns the general-used commercial instruments, requires the comprehensive analysis of the cardi-osignals. The differences between various cardiosignal charts obtained using modern hardware and charts obtained using the earlier known hardware have recently become more apparent[22] [13]. Nevertheless, the issue of reverification of the information criteria applicable for the earlier used signals has not been yet discussed in the scientific literature. The formal transfer of the information criteria obtained using the instruments

21   Cardiology Manual / N.A.Manak, V.M.Alkhimovich, V.N.Gaiduk & et al. Mn.: "Belarus", 2003, 624 pages
22   In the same reference.

of the previous generation into the modern information systems gives rise to the various contradictions in terms of the information criteria interpretation when diagnosing any disease. From our point of view, this problem is caused by the lack of theoretical and practical proof of the information criteria formation in recorded signals.

However, the attempts to describe the biophysical processes that determine the hemodynamics mechanism are always in progress[23] [3]. But only technicians deal with this work. The scientific and economic potentials of the technical sphere exceed the medical potential greatly. The application of the hydrodynamics theory for description of hemodynamics allowed to explain a lot of the biophysical processes that form the blood circulation[24] [3]. But no one has still managed to explain the relationship between various factors that enable the minimum energy costs of the real hemodynamic process.

23   Caro C, Padley T., Shroter R., Sid W. Blood Circulation Mechanics, M.: "Mir", 1981. P.624.
24   In the same reference.

On this background, in 1980, G. Poedint-sev and O. Voronova pro-posed a theory of high fluidity of liquid superfluidity[25] [14]. While studying the technical problems of hydro- and aerodynamics in a professional manner, they managed to simulate liquid superfluidity conditions in the rigid pipes. For this purpose, the dedicated hydrau-lic pulsators were used. It was recorded that the power consumption that enables the high fluidity of liquid was reduced almost tenfold[26] [14] as compared to the power consumption required for laminar process. But in this case, the given conditions became very similar to the mode of the pulsed arterial pressure in the cardiovascular system. The further studies showed that physical processes, which enable the liquid superfluidity in the rigid pipes, are the same as the physical processes, which enable the blood circulatory system hemo-dynamics. The developed mathematical tools technique, which describes superfluidity conditions, was used when describing the hemodynamics processes.

Let us consider the milestones of the hemodynamics theory pro-posed by G. Poed-intsev and O. Voronova.

The authors have found out that the following processes take place in the rigid pipes at the instant of the liquid flow initiation under certain conditions: when, due to the initial static pressure differen-tial in the pipe, the liquid particles displacement along the pipe axis is initiated, the traveling wave occurs in the boundary layer. The wave front is directed toward the pipe axis[27] [14] (Fig. 5).

The amplitude of this wave depends on the pipe diameter, liquid viscosity and initial differential of the pressure at the ends of the pipe. During the motion of liquid particles along the pipe axis, the above-men-tione d traveling wave will cont i nue spre ad-ing to the center point of the pipe axis. During this complex process, the

---

25   Voronova O.K. Development of Models & Algorithms of Automated Transport Function of the Cardiovascular System. Doctorate Thesis. Prepared by Mrs.O.K.Voronova, Ph.D., Voronezh, VGTU, 1995, 155 pages
26   In the same reference P. 40.
27   Voronova O.K. Development of Models & Algorithms of Automated Transport Function of the Cardiovascular System. Doctorate Thesis. Prepared by Mrs.O.K.Voronova, Ph.D., Voronezh, VGTU, 1995, P. 51-52.

length of the traveling wave is continuously increasing and finally, the wave contour in the pipe full cross-section assumes a shape of the parabolic curve that corresponds to the laminar flow (Fig. 6).

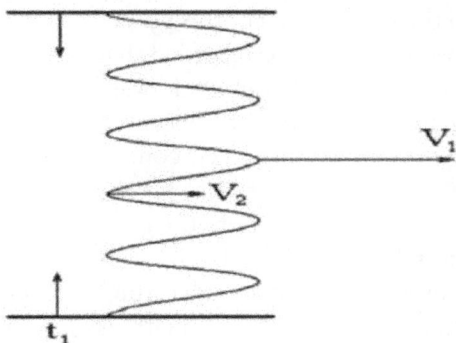

Fig. 5. Traveling wave generation at the instant of liquid motion initiation
(by G. Poedintsev and 0. Voronova)

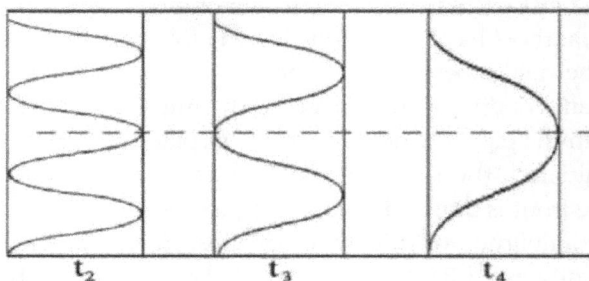

Fig. 6. Change of the traveling wave contour
(by G. M. Poedintsev and O. K. Voronova)

As the authors note, during this very short time period (i. e. from the moment when the particles begin to move till the moment when the laminar flow is shaped), the liquid moves in the economic energy mode thanks to minimum friction between the liquid particles.

This friction will increase sharply (due to the edge effect) as soon as the laminar flow is shaped.

To distinguish from turbulent and laminar flows, the authors called this phenomenon "the third fluidity condition".

One further peculiarity of the third fluidity condition is the following phenomenon: if the liquid contains the suspended particles, which are similar to the blood cells, then in the course of development of the described wave process, the particles will be concentrated in the wave maxima, while the pure portion of the liquid will be concentrated in the wave minima[28] [14]. During the motion of the liquid of such structure along the pipe axis, the velocity of the concentric layers with the particles will be two times higher than the velocity of these layers without the particles. The velocity vectors will be parallel to the flow axis. And that is the very condition of superfluidity when the friction between the layers and pipe walls is reduced.

The shaped concentric liquid structure exists in the rigid pipes for the very short period. In the living organisms, this structure is kept thanks to the vascular distension. Moreover, the rapid vascular distention in the initial stage is followed by slow contraction of vessel. This condition exists due to arterial pressure fluctuation the shape of which is shown in Fig. 7:

---

28   In the same reference. P. 51.

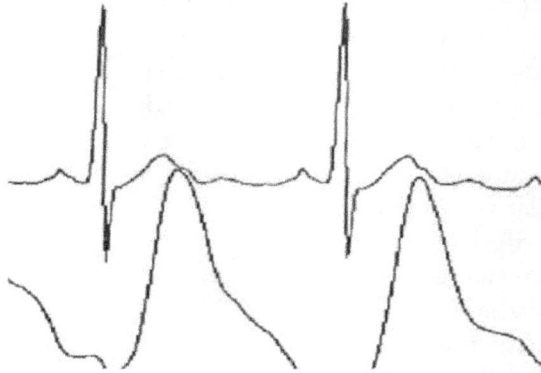

Fig. 7. Shape of arterial pressure wave recorded using rheographic device. ECG trace is shown in synchronism with RG curve. Arterial pressure fluctuations ensure the blood superfluidity

The authors who have described the third liquid superfluidity conditions have proposed the mathematical equations to calculate all flow modes. These very equations were also proposed by the authors to calculate the cardiovascular system hemodynamic parameters[29] [15]. The basic original equations used to calculate hemodynamic parameters are as follows[30] [15]:

$$r_t = r_0 \left(\frac{t}{t_0}\right)^{1/5} \qquad (3)$$

$$W_t = W_0 \left(\frac{t_0}{t}\right)^{2/5} \qquad (4)$$

where r_t is reference radius of the pipe widening;
r_0 - is basic radius (at $t = t_0$);
t – is current time;

---

[29]   Patent No.94031904 (RF). Method of Determination of the Functional Status of the Left Sections of the Heart & their Associated Large Blood Vessels. Authors: Poyedintsev G.M., Voronova O.K.
[30]   In the same reference.

36

t_0 - is time required to accelerate the flow up to the maximum speed at the instant of impulse;

W_t - is the current value of the liquid motion speed;

W_0 - is the maximum value of the speed at the instant of impulse (at $t = t_0$).

As it was mentioned earlier, phases of cardiac activity are basic characteristics of hemodynamics. During one cardiac cycle, the shape of the heart is changes about ten times. This corresponds to the phases of the heart work[31] [7].

The question arises why the developed mathematical tools technique, which claims to be the master technique in the hemodynamics theory, has not become common use. The fact is that the authors of this technique have used the polycar-diographical method to measure the duration of the cardiac cycle phases. As noted above, this method gives significant errors. As a rule, the papers published at the earlier stages of research described the polycardiographical method but did not provide signals, which were actually recorded. They included just drawings of signals that are gray-shaded. For an example refer to Fig. 2.

Fig. 8 shows the kinetocardiography signals obtained in a number of derivations and electrocardiosignals which were actually recorded on paper.

It goes without saying that the information criteria of the cardiac cycle phases manifested in the kinetocardiogram signal can be determined only by the qualified specialist who has skill and experience in using this diagnostic technique. There is no point in speaking of measurement accuracy, because the measurements are taken manually on paper using a cali-per and a ruler. Only after these measurements are completed, the obtained results are entered into computer manually and are processed by the special software.

It is quite natural, that such approach, which does not comprise the total set of the above-mentioned simulation theory, the theory

---

31   Andreyev L.B., Andreyeva N.B. Cinetocardiography. Rostov-on-Don, Published by the Rostov State University, 1971, 308 pages.

of the biophysical processes and the theoretical and practical proof, did not allow to computerize the process of the cardiac cycle phases interpretation.

Thus, the theory paradigm in the cardiology came up to the necessity to be totally updated based on new knowledge that not only allows to raise the question of diagnostic signals measurement accuracy but also determines the ways for goal achievement.

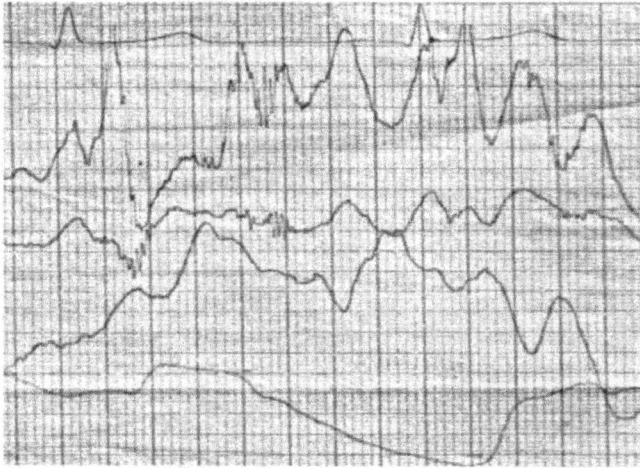

Fig. 8. ECG signals and kinetocardiogram signals (obtained in various derivations), which were simultaneously recorded on paper

1.4. Prospects of hemodynamics research based on electrocardiosignals

As it has been pointed out, the cardiovascular system examination based on cardiosignals is the most promising method for diagnostics. It is largely due to fact that the functional diagnostics of the heart using ECG recording is available to largest majority of population. But the first reason for this is successful computerization of ECG recording procedure which was initially intended to research phasic structure of cardiac cycle. It is the opinion of many experts, that ECG multichannel recording makes it to some extent possible to estimate phases of heart work.

Cardiac cycle is described in different ways in scientific literature. In earlier editions of papers written by L. Andreyev and N. Andreyeva, for instance, in the book "Kinetocardiography"[32] [7], we can find the cardiac cycle block diagram (Fig. 9).

One can see that cardiac cycle falls into systole and diastole. Systole falls into tension period and ejection period. Diastole falls into relaxation period and filling period. Each period falls into two more phases. Only filling period falls into three phases. So, cardiac cycle comprises nine phases.

The same edition is provided with a figure, which presents various graphic signals used in polycardiography; this figure illustrates signal synchronous variation, that is rheogram, phonocardiogram and cardiogram (Fig. 10)[33] [7]. It is the opinion of the authors that it is namely the comparative analysis of these three signals that allows to outline informative criteria for cardiac cycle duration recording. However, the authors do not refer to A. Guy ton in their book. The recent publications of other authors also provide the illustrations of heart phases. For example, in "Cardiology Manual" issued in 2002 we can see the figure describing the phases of heart work (Fig. 11).

---

32   Andreyev L.B., Andreteva N.B. Cinetocardiography. Rostov-on-Don, Published by the Rostov State University, 1971. 308 pages.
33   In the same reference. - P. 18.

In the caption, the name of A. Guyton (1981) is mentioned as the author of this research. In fact, two figures are identical. In spite of the thirty years interval between publishing of these figures by independent authors, one may have an impression that research theory has fallen into stagnation, since no feasible signals, which have been recorded, have been published over this period.

In the caption, the name of A. Guyton (1981) is mentioned as the author of this research. In fact, two figures are identical. In spite of the thirty years interval between publishing of these figures by independent authors, one may have an impression that research theory has fallen into stagnation, since no feasible signals, which have been recorded, have been published over this period. Only sketches, which were so adequately made by the drawers based on description furnished by experts, have been published in many books and papers within the period of more than twenty-five years[34] [3]. We have not found a publication which provides the analysis of pathologic condition signals sketched by the drawers.

34   Caro C, Padley T., Shroter R., Sid W. Blood Circulation Mechanics, M.: "Mir", 1981. P.231.

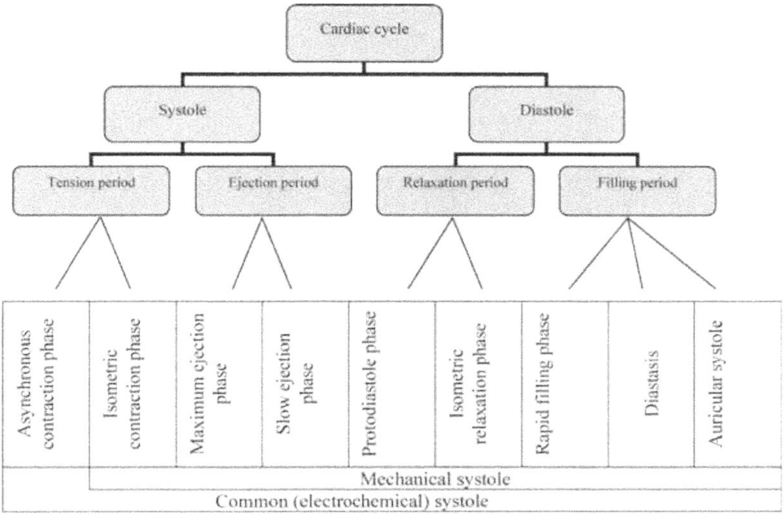

Fig. 9. Cardiac cycle ("Kinetocardiography", page 16, issued in 1971)

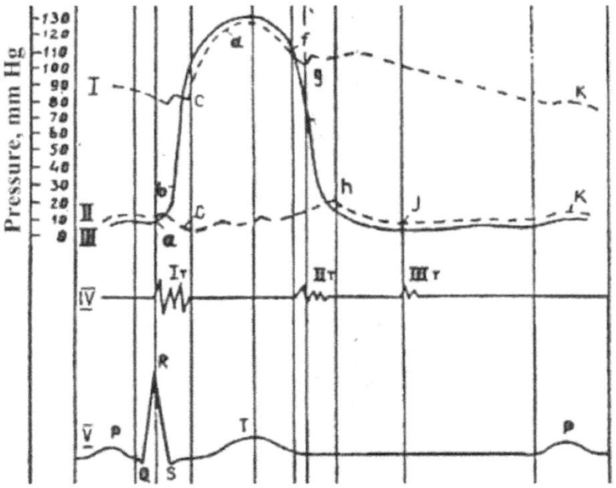

Fig. 10. Synchronous recording of rheogram, phonocardiogram and ECG signals
("Kinetocardiography", page 8, 1971 edition)

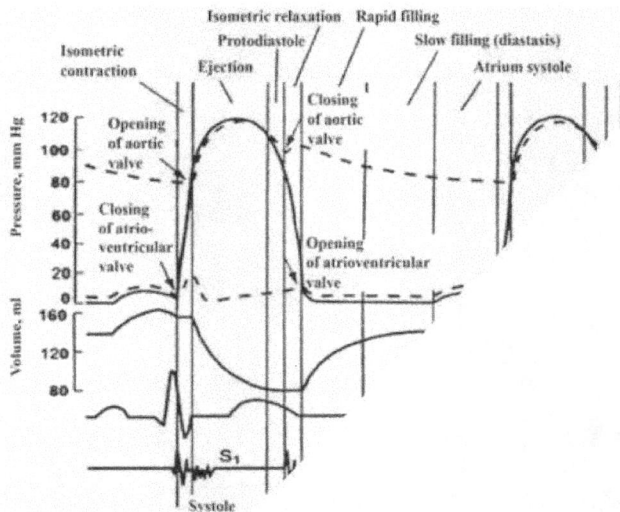

Fig. 11. Synchronous recording of rheogram, phonocardiogram (PCG) and ECG signals ("Cardiology Manual", page 30, issued in 2002)

Thus, we come across the existence of the problem of authenticity and reliability of cardiosignals recording. This fact hinders greatly the progress of phase analysis theory and practical application of analysis.

Against this background, the paper of V. Karpman is very remarkable[35] [16]. He was among the first to propose the application of one type of signals, i.e. electrocardiogram, for cardiac cycle phases analysis. Theoretically, such approach is well-founded, since ECG is used in the poly-cardiographic method as one of the main informative signals. Moreover, it is important to say that the multichannel recording of ECG provides reliable information based on which the determination of cardiac cycle duration[36] [16] can be evaluated.

Usage of only one signal minimizes errors introduced via other recording channels, via kinetocardiographic channel in particular.

35   Karpman V.L. Phase Analysis of the Heart Function. M., "Medicina", 1965. 328 pages.
36   In the same reference.

Thus, only parameters of the amplification path, which conditions the recorded signal up to the level which ensures AD conversion accuracy, may be a source of measurement error. Dynamic range and pass band are among those parameters.

At first sight, the isolation of ECG with a particular dynamic range of signal-to-noise ratio might be treated as a mere engineering task. In terms of radiotechnics, this task can be solved quite easily. By changing the electronic filter passband, we can obtain a signal-to-noise dynamic range of more than 40 dB. This is the very level, which provides further analog signal-to-digital code conversion using twelve-digit analog-to-digital converter with level of error equal to 1%. If we expand the dynamic range, we can use the analog-to-digital converter with a higher resolution.

But we should not ignore the probability of distortion due to signal filtering. The issue of medicobiologic signals recording errors is covered in literature only in general terms.

So, Ye. Popechitelev and N. Korenevsky in their tutorial for bioengineers treat this problem in the light of a combined method for electrophysiological methods errors research[37] [8]. Among other things they point out the following: "Analysis of the electrode-dermal contact impedance and its effect on signal amplitude transmission accuracy does not cover all the problems associated with the research of electrophysiological information read-out errors. It is known that various random factors affect the accuracy of signals recording and confidence level of interpretation of electrophysiological research results. However, not all factors can be monitored and controlled"[38] [8]

The authors propose a combined method of signal measurement error research: "1. Determination of the main factors which can cause errors. To main factors belong well-interpreted measuring system parameters through which the effect of all other random factors can

37   Popetchitelev Y.P., Korenevsky N.A. Electrophysiological &Photometric medical Equipment. M., "Vysshaya Shkola", 2002, P.75-77.
38   Popetchitelyev Ye.P., Korenevsky N.A. Electrophysiological & Photometric Medical Equipment. M., „Vysshaya Shkola", 2002, P.76.

be manifested. These main factors must be quantified or they must refer to some law of variation"[39] [8].

The quantitative record of the main factors which are error sources is based on statistical analysis. It is at present far from being able to determine the information criteria, which can be distorted by the signal conversion principle itself. In order to identify the true information, which is contained in the recorded signal, it is necessary to know the law of absolute (not statistical) error variation, which is different from the ideal model serving as a reference.

Lately, the scientific papers have been published in which the authors apply the principles of signal mathematical graphical differentiation[40] [17]. V. Fatenkov, in particular, recommends to determine the cardiosignal information criteria using the curve of second differential coefficient for function, which appears to be the same as the curve of the cardiosignal recorded from the body. Here one can also find the examples of normal/pathology signals. The peculiarity of the suggested method is the application of the information criteria for the second differential coefficient curve.

However, ECG signal is not differentiated. Differentiation applied to kinetocardiogram signal (here we mean the kinetocardiogram recorded simultaneously with the rheogram and the ECG). The usage of kinetocardio-grams itself implies significant errors in measurements. That is why the authors cannot describe the theoretical model of biophysical processes which form the information criteria in the method suggested by them.The opinion of A. Kramarenko who used ECG differentiation is worth of being considered. „The amplitude of the differentiated ECG (i.e. filtered ECG with the band limited from below) correlates with spectral component displacement value. To estimate such amplitude means to apply a primitive method which has many errors caused by the variability of

---

39   In the same reference. P. 77.
40   Fatenkov V.N. Biomechanics of the Heart, Phase Structure of Heart Cycle. Samara, The Samara State Medical University, 1998, 16 pages.

all ECG amplitude characteristics, but this method really works and is efficient"[41] [18].

ECG differential analysis has also been studied by N.V. Bogdanova in her thesis. N. Bogdanova remarks: "... the algorithm of ECG-type curves differentiation on spline-approximation basis substantially improves the accuracy of ECG analysis"[42] [19].

The perspective of application of mathematical differentiation in ECG analysis is quite obvious since it is a really functional and efficient means which improves the accuracy of the ECG processing. But the given perspective hangs up at the stage of laboratory tests and does not become a common practical use. The reason for this is the obtained qualitative analysis of only some of the phasic characteristics of heart work instead of quantitative estimation of hemodynamic parameters. That is the reason why the diagnostic based on ECG analysis is less efficient than the diagnostic based on ultrasonic cardiography. However, the determination of the essence of the biophysical processes which take place in each of the phases of cardiac cycle as well as the computation of values of phase duration for the normal heart physiology is definitely an achievement in the phase analysis obtained using polycardiography. The data are presented in Table 1. This table specifies the biophysics of phases, presented in Fig. 9. The table is valuable because it provides the precise duration of each phase in the norm. Other papers also cover the ranges of normal duration of cardiac cycle phases[43],[44] [13].

It has been pointed out already that the application of only ECG recording is sufficient for the determination of phases duration. Hardware and software can record and process ECG quite effectively.

---

41   Kramarenko A.V. RED as Universal Indication of an Affection of the Cardiovascular System [electronic resource]: Access: http://www.dx-telemedicine. eom/rus/publications /hubble/hubble Ol.htm. Screen title.

42   Bogdanova N.V. Development of Algorithms and Application Software for Cardiologist-Reseaercher's PC-Workstation. Thesis prepared by 2001.06.27.

43   Cardiology Manual / N.A.Manak, V.M.Alkhimovich, V.N.Gaiduk & et al. Mn.: "Belarus", 2003, P. 32.

44   In the same reference. P. 46.

ECG signal recording procedure is also quite easy. Minimal costs have ensured its wide practical application.

However, phase analysis is primarily the determination of parameters of heart mechanical work which provides organism hemodynamics during its contraction. The following questions arise: how are mechanical oscillations and ECG electrical potential interrelated? why is it so that nowadays only determination of time interval of phases is used in practice?

Let's consider the amount of information obtained from value of phases duration provided by ECG using Table 1 (Fig. 12)[45] [13].

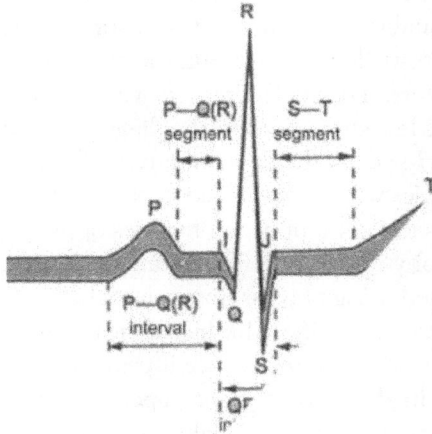

Fig. 12. Established information criteria for determination of cardiac cycle phases duration based on ECG (N. Manak and co-authors, 2003)

---

45   In the same reference.

Table 1.

Cardiac cycle phases, their essence and duration.

Phase	Beginning of phase	End of phase	Duration, s
1 – isometric contraction	Beginning of ventricular contraction	Opening of semilunar valves	0.05
2 – ejection	Opening of semilunar valves	Beginning of ventricular relaxation	0.22
3 – isometric relaxation	Beginning of ventricular relaxation	Opening of atrioven-tricular valves	0.12
4 – rapid filling	Opening of atrioventricular valves	Slow ventricular blood flow	0.11
5 – slow filling (diastasis)	Slow blood flow from auricles to ventricles	Beginning of auricular contraction	0.19
6 – filling due to auricular systole	Beginning of auricular contraction	Beginning of ventricular contraction	0.11

The following intervals: P - Q(R), QRS, S -T, Q - T and segments: P - Q(R), S - T are shown in figure. If we compare these phases with Table 1, we shall obtain the following correspondence:
1 - isometric contraction QS.
2 - ejection S-T.
3 - isovolumetric relaxation part of wave T, from inflection point to T end.
4 - rapid filling from T wave end to U wave end.
5 - slow filling (diastasis): from end of wave U till peak of wave P.
6 - filling due to atrium systole P - Q.
Cardiac cycle phases shown by diagram in Fig. 9 correspond to the following duration of phases (with respect to the revised names of phases):
1 - isovolumetric contraction RS
2 - ejection S-U
3 - isovolumetric relaxation part of wave, from inflection point to T end.
4 - rapid filling from T wave end to U wave end.
5 - slow filling (diastasis) from end of wave U to peak of wave P.
6 - filling due to atrium systoleP-R.

When considering both correspondences, the difference in under-standing of heart phase work biophysical processes becomes obvi-ous. The term "electrocardiography of high resolution»[46] [20] has been suggested lately when studying this problem. Its essence is in the capability to record HF oscillations, of 20 to 50 Hz with ampli-tude of 5 to 20 |JV, which occur after QRS complex[47] [20].
The method of recording is related to recording of 500 cardiac cy-cles, which are subjected to time averaging and spectral analysis. As a result, we obtained the specified oscillations which are designated as "later ventricular potentials"[48] [20].

---

46   Ivanov G.G. High-Resolution Electrocardiography. M., "Triada-X", 1999, 280 pages.
47   In the same reference. P. 19.
48   In the same reference. P. 19-24.

Research shows the "later ventricular potentials" relation to the sudden cardiac death[49] [20]. Their occurrence seems to affect the tachycardia mechanism which ends in cardiac arrest emergency[50] [20].

Those facts presented above prove the importance of the further research of phase processes of cardiac activity. Successful solution of this problem is directly related to solution of the problem of cardiovascular system modeling.

---

49   In the same reference. P. 182.
50   In the same reference. P. 244.

2. Methodology for electrocardiosignals research based on graphic mathematical differentiation

2.1. Evaluation of medicobiological signals distortion during recording

The research of medicobiological signals is made by various ways. The variety of signals does not allow their definite systematization. However, the common purpose of all research activities is to reveal diagnostic criteria which allow identification of specific parameter of body functioning.

The metrology aspects of measurement and registration of medicobiological signals, and assessment of reliability of criteria to be used for defining the functional parameter are the major issues of diagnostics.

The metrology aspects have been covered in the first section of the present paper. But it is necessary to make a note about a number of problems in this section too. If the error is taken to be a difference between the observed and reference values then at least two prerequisites are required. The existence of the proper reference value is the first pre-requisite, the confidence in reliability and adequacy of data obtained during assessment of diagnostic criterion.

In general, the model representing the basic properties of the object is taken as a reference. In technical research, the reference values are strictly determined and there is always an opportunity to measure the required parameter with high degree of rel iabil ity, wh i le in me dici ne no mo del of human body is available today. The measurements of physiological parameters are effected by direct or indirect methods.

The devices which make measuring by direct methods can be verified through metrological testing. As an example we can take the devices which measure blood pressure by a direct method. It seems, that such device authentically measures the arterial pressure in major blood vessels with required accuracy[51] [5]. It can be checked up with use of the model which builds up pressure on the sensor. The model reflects basic property which is measured, i.e. pressure. The sensor (transducer) can be calibrated directly prior to taking the measurements. Calibration is mainly a process of normalization of the gain-transfer performance of the device.

However, the influence of the sensor (transducer) on the arterial pressure to be measured is not considered in this example. Account must be taken of degree of reliability of measured value of arterial pressure bearing in mind the influence of the sensor on a nature of arterial pressure formation. A. Guyton has described in detail this problem while investigating the minute volume[52] [21]. He has noted, that the sensors used for direct measurement of hemodynamic parameters can distort the true values[53] [21]. Therefore a problem of reliability in measurement of medico-biological parameters by direct methods is not completely solved yet.

More complicated problem arises with use of indirect methods of measurement. Indirect measurement of arterial pressure can serve as an example. The only one indirect method has been used for more than hundred years. It is the method of blood pressure measurement using the cuff[54] [5].

Two aspects are to be mentioned when considering the cuff method. On the one hand, there is a possibility of calibration of manometer used for registration of arterial pressure. On the other hand, the generally accepted criterion for registration of equality of the occlu-

---

51   Medical Electronic Equipment for Public Health Service, Transl. from English into Russian, M.Arditti, F.Waybell et al., Translation edited by R.I.Utyamysheva, M., "Radio I Svyaz", 1981. P.147 -152.
52   Gaiton A. The Minute Heart Volume & its Regulation. M., "Medicina", 1969, 472 pages.
53   In the same reference.
54   Medical Electronic Equipment for Public Health Service, Transl. from English into Russian, M.Arditti, F.Waybell et al., Translation edited by R.I.Utyamysheva, M., "Radio I Svyaz", 1981.

sive and arterial pressure is used, but this criterion is not positively proved in theory and is not provided in terms of metrology.

In electronic devices used for measurement of arterial pressure, so-called os-cillometrical method is used[55] [22]. As it was mentioned in the first section, the amplitude detector with fixed operating threshold of the comparator is used for revealing the criteria of pressure equality moments[56] [23]. This criterion for arterial pressure measurement is established based on statistical data. It imposes a lot of restrictions on the measurement technique.

Russian scientists set forth a model of biophysical processes of occlusive blood flow with the purpose of correction of a theoretical substantiation of criteria used in indirect oscillometrical method of meas-urement[57] [11]. The shape of recorded os-cillogram was investigated using double differentiation. The influence of electronic filter passband variations on the shape of oscillogram was estimated. The method of double differentiation enabled to establish that the shape of oscillogram reflects the process of arterial pressure direct and return (reflected from point of occlusion) waves interference. The criteria for measurement of arterial pressure were identified by simulating the interference on a pulsing flow in elastic pipes. Figure 13 shows the synchronous records of signals which characterize the interference of arterial pressure waves.

55   Savitsky N.N. Biophysical Principles of Blood Circulation Clinical Methods of Researches in Hemodynamics. L., Medicina", 1974, 312 pages.
56   Reference Book. MOTOROLA Sensors.
57   Rudenko M.Yu., Alexeyev V.B., Matsyuk S.A. Biophysical henomena in Blood Circulation in Indirect Measuring Arterial ressures & Evaluation of the Relevant Instrumentation // Medtechnica". 1986 No.5 P.26-35.

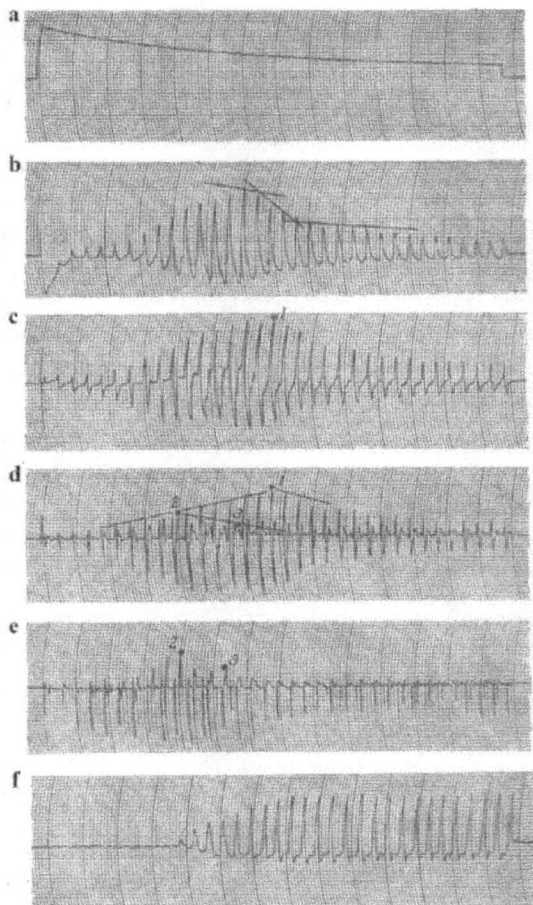

Fig. 13. Oscillography method for indirect measurement of arterial pressure. The method is based on mathematical graphical differentiation:

a) pressure in occlusion cuff;

b) tachooscillogram of arterial pressure oscillations in area of vessel occlusion by cuff;

c) first-order derivative of tachooscillogram,

d) second derivative,

e)informational part of second derivative for systolic pressure measurement,

f) photoplethysmogram of vessel distal of occlusion area

In 1984 the given method was implemented in the electronic device which was certified for series production by regulatory agencies after thorough clinical tests. The reliability of offered criteria for measurement of arterial pressure was the main point for certification.

It is necessary to note, that the passband variation distorted the criteria which characterize the pressure equality moments. It became possible to identify measured parameters only after the comprehensive reasoning and resolving the theory of biophysical phenomena in oc-clusive blood flow and through construction of the visual model representing the basic properties of compressed artery.

To verify this idea, the ultrasonogra-phy method was used for investigation of occlusive blood flow in areas which are distal and proximal to area of registration of oscillations. The results were compared with signals obtained from the model[58] [11].

Oscillogram was subjected to double differentiation. It was the usage of derivatives that enable to describe the biophysics of process and, as a final result, to develop an essentially new device.

The research has proved once again, that the characteristics of any signal, including a medicobiological signal, can be revealed with the help of mathematical differentiation. It was also verified, that the first derivative shows change of energy which shapes a medicobiological signal, and the second derivative shows the change of the amplitude characteristics of a signal when there is an interaction with the external factors. The external factors can be caused not only by the exercise stress, but also by pathology of some kind. The described method of the estimation of medicobiologi-cal signals with use of mathematical differentiation has a great potential.

It is necessary to pay attention to a problem of signal distortion by measuring systems.

Any electronic device necessitates the selection of a signal in a required frequency band. For cardiosignals recording, 0.5 to 20 Hz frequency band is standard.

---

58   In the same reference.

Operational amplifiers are used for cardiosignals amplification in the above-mentioned frequency band. The equation of input and output signals relationship is as follows[59] [24]:

$$U_{output} = -R_{feedback} * C * d\frac{U_{input}}{dt},\qquad(5)$$

One can see from the equation that actually any signal is a derivative signal, as the right hand side of the equation contains a derivative function alongside with values of feedback elements. Derivative depends on the amplitude of the input signal.

As for the sinusoidal signal, its derivative will be represented by a cosinusoid leading the initial signal by 90 degrees. As one can see in figure 14, points of ex-tremum of sinusoid (which is a function in general case) coincide with points of curve inflection. In electronic devices, the differentiation is implemented through calculation of filters passband. The filtering ensures the accuracy of differentiation not throughout entire passband, but only at a level not lower then 0.707 of maximal signal amplitude for frequencies which lie in area from the bottom boundary to top boundary of cutoff.

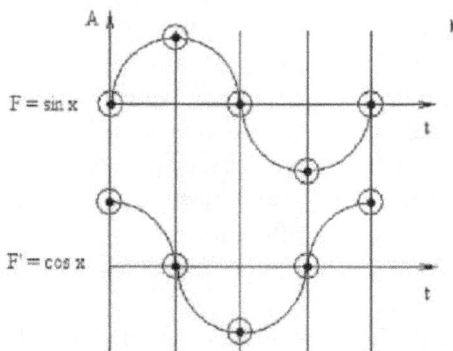

Fig. 14. Sinusoidal function and its first-order derivative

59   Rutkovsky J. Integral Operational Amplifiers. M., "Mir", 1978, P. 180.

The frequencies above upper cutoff frequency are subjected to integration. The equation for these harmonics has the following form:

$$U_{вых} = -\frac{1}{RC}U_r dt$$

The integration results in distortion of harmonics, whose frequencies are higher than upper cutoff frequency. The distortion increases as the input signal amplitude increases. This distortion will manifest itself by pulling of leading edge of a pulse. On ECG trace it is shown by smoothing the QRS complex.

The real medico-biological signals are complex spectral signals, they consist of the sum of sinusoidal signals of various frequency. If data signals are made of harmonics spaced in frequency and exceeding cutoff frequencies, then the conditions of strict differentiation are not met strictly and integration function will bring some essential distortions. Some harmonics will correspond to second-order derivative while the other harmonics will rank below the first-order derivative. In this case the information will be distorted.

It is impossible to increase the pass band to infinity. The useful signal starts being distorted by noise, mostly by flicker noise[60] [24] of 1/f kind. The described problems can not be solved through computer processing since the signal should be conditioned prior to its analog-to-digital conversion by amplification in frequency band. So, a compromise has to be reached when choosing between the amplifier passband and signal-to-noise ratio which should not be less than 40 dB.

The nature of noise of 1/f type depends to some extent on intensity of current consumed by the device, and on configuration of grounding busses. At present, we can use a micropower AD converters and controllers, and infra-red transmission channels to ensure the optimal electrical isolation and to enhance the signal-to-noise ratio to level more than 50 dB.

---

60   In the same reference. P. 128.

But still it is necessary to make sure that the above mentioned conditions are sufficient to obtain non-distorted data required for reliable diagnostics criteria. Practical prospects of a developed method depend on this condition.

There is where we have the problems related to metrology issues. As noted above, the solution can be found through development of more reliable biophysical process theory. These processes might be simulated and compared with results obtained using the instruments which are under development process. Such approach will enable to find the metrologically guaranteed methods of measurement, to evaluate the measurement error and to ensure the required accuracy of measurement. Taking into account the lack of metrological provision in practical cardiology at present, the given approach will not only refine the existing methods, but will also expand opportunities of practical diagnostics which is based on ECG tracing.

The modeling allows application of general properties of phenomenon to an individual object. Naturally, it establishes limits for parameter measurement accuracy, sufficient for the given level. The theory should not have "blind spots". Each assumption must be proved theoretically and checked by practice under various conditions.

Research of hemodynamic processes in blood flow, particularly the analysis of cardiac cycle phases, should be effected taking into account the above-mentioned issues of medical information systems modeling and design.

2.2. About blood vessel sanguimotion conditions

(This section is written by G. Poedintsev.
It was planned to incorporate more information into this section and to
issue the revised version of the given work.
But the fate has ordered otherwise.
Based on the corporate authors decision,
we represent this section without any revisions.).

The development of medical science goes hand in hand with the search for physiological processes, the measured parameters of which would enable to obtain possibly great amount of information on the general state of the human organism. The researches of the latest three decades have shown that one of the above-mentioned processes is the sanguimotion in blood vessels[61],[62] [25, 26], A. L. Chizhevskiy has put the diagnostic and informative importance of the flowing blood in the following words: "There seems to be no one disease, even the minor one, that would not affect immediately the space pattern of the blood since blood is a "mirror" that reflects the organism state. And vice versa, any deviation of blood particle distribution from the norm affects the general state of all human organism"[63] [25]. He has found out that "in case of normal or near-normal concentration, the erythrocytes shall form inside the bloodstream well-shaped assemblies like rouleaus that are concentrically arranged in the orthogonal cut"[64] [25].

This important phenomenon of the flowing blood can not be explained in terms of the modern theory of blood circulation hydrodynamics. The fact is that, in opinion of most investigators, the blood

---

61   Tchizhevsky A.L. Structural Analysis of Moving Blood. M., Academy of Sciences of the USSR, 1959, P.3.
62   Tchizhevsky A.L. Electrical & Magnetic Properties f Erythrocytes. Kiev, "Naukova Dumka", 1973.
63   Ref. to 61.
64   Ref. to 61.

flow is laminar in the whole cardio-vascular system, except for the heart and aortic ostium65 [27]. Moreover, it is assumed that Poiseuille's law[65],[66],[67],[68],[69],[70][28, 3, 29, 30, 31], which is presented below, can be applied to the laminar blood flow conditions:

$$Q = \frac{\pi r^4 \Delta p}{8 \mu 1}$$ (7)

where:
Q - is flow rate;
r - is pipe radius;
Δp - is pressure gradient;
μ - is viscosity factor;
l - is pipe length.

In addition to the above-mentioned law, Poiseuille flow has the following features: the parabolic velocity distribution, lack of the static pressure lateral gradient and, as a consequence, lack of the hydrodynamic forces required for the blood flow pattern formation, i.e. required to form the rouleaus.
Both the phenomenon of blood flow pattern formation and some experimental facts give rise to doubts about the applicability of Poiseuille's law to the blood vessel san-guimotion in vivo. For example, enormous losses due to friction if sanguimotion is effected based on Poiseuille law.

---

65   Regirer SA. Some Issues of Blood Circulation Hydrodynamics. In a Miscellany of Translations: Blood Circulation Hydrodynamics. M., "Mir", 1971, P.242-258.
66   Folkov B., Nil E., Blood Circulation. M., „Medicina", 1976, P. 19.
67   Ref. to 3.
68   Cardiology Manual, under Editorship by Ye.I.Tchazov. M., "Medicina", 1982 Vol. 1 P.195.
69   Johnsov P., Peripheral Blood Circulation, M., "Medicina", 1982 P.136.
70   Remizov A.N. Course in Physics, Electronics & Cybernetics. M., "Vysshaya Shkola", 1982 P.96.

According to the principle of optimality, which is inherent and organic in wild-life[71] [32], the processes that take place in the living systems have the highest coefficients of efficiency. Therefore, probably the blood vessels also have high hydraulic efficiency.

The hydraulic efficiency of a pipe, as well as of any blood vessel, shows the fluid flow dynamic head-to-total head ratio, with total head being equal to the sum of dynamic head plus friction resistance head[72] [33].

$$\eta = \frac{1}{1+\lambda\frac{1}{d}}, \qquad (8)$$

where
η- is hydraulic efficiency of pipe;
λ - is hydraulic friction factor;
l - is pipe length;
d - is pipe diameter.

In case of laminar flow, λ factor is a function of Reynolds number

$$\lambda = \frac{64}{Re}, \qquad (9)$$

where Re - is Reynolds number.

71   Rosen P. The Principle of Optimization in Biology. M., "Mir", 1969 P.71.
72   Altshul A.D. Losses due to Friction in Piping. M.,"Stroyizdat", 1963 P.19.

This formula implies that the less Reynolds number is the higher is A factor and the less is pipe hydraulic efficiency. For example, would the pipe have the same parameters as the arteriole[73] [34] (i. e. l = 2 mm, d = 0.02 mm, Re = 0.02 mm), the value of pipe hydraulic efficiency would be no higher than $3*10^{-6}$. Due to the elasticity of pipe walls, the hydraulic efficiency of the pipe may be increased, but not more than by an order of magnitude.

The given example shows that the concept of Poiseuille conditions of blood vessel sunguimotion in vivo does not correspond to the principle of optimality in biology.

It has been found out experimentally by means of high-speed photography of the flowing blood near vessels margins[74] [35], that there are transverse forces in the real stream of blood. These forces provide erythrocytes deflection from the rectilinear longitudinal trajectory. The recognition of availability of such irregular distortions of blood flow in all vessels, except for the true capillary tubes, allows to assign the blood flow conditions in these vessels rather to pseudo-turbulent flow[75] [36] than to Poiseuille flow.

The pseudoturbulent flow differs from the common turbulent flow in following: the intensity of the irregular motion is determined not by Reynolds number, but by the level of shear rates in the middle flow. The typical scale of vortexes occurred in such a flow is equal to the erythrocyte size and does not depend on Reynolds number[76] [27].

However, the pseudoturbulent flow, as well as the turbulent flow, has not been yet rigorously defined in terms of mathematics. Therefore, replacement of the concept of Poiseuille's conditions of blood vessel sunguimotion in vivo with the concept of pseudoturbulent flow conditions does not provide a prospect for establishment of new laws for blood flow. This replacement only shows once again

73   Whitmore R.L. Rheology of the Circulation. Oxford: Pergamon Press, 1968.
74   BlochE.H.-Amer. J.Anat., 1962, 110, M 2, 125.
75   Levtov V.A., Regirer SA., Shadrina N.Kh. Blood Rheology. M., „Medicina", 1982. P.188.
76   Ref. to 65. Shlichting G. Theory of Boundary Layer. M., IL, 1956, P.57.

that the solution of in vivo hemorheology problems completely depends on the solution of hydrodynamics problems, among which the
problem of obtaining the general complete solution of the viscous
fluid flow equations[77] [37] is of most vital importance.

Thus, let us put aside the analysis of the blood vessel sunguimotion
and carefully analyze the classical experiment that demonstrates
Newtonian fluid flow in the long straight horizontal pipe. Let us
clarify which aspects of the above-mentioned flow are known and
which aspects are still unknown.

The facility (Fig. 15a) used to carry out this experiment comprises a
long straight horizontal pipe 1. This pipe is of circular cross section.
The diameter of the section is equal to d. Both inlet and outlet of
the pipe are connected with rather big reservoirs 2 and 3, so the
pressure differential at the ends of the pipe can be kept constant for
a long period, thus, the stationary flow is ensured. A quick-acting
gate is mounted into the pipe to facilitate instantaneous opening of
the pipe flow area.

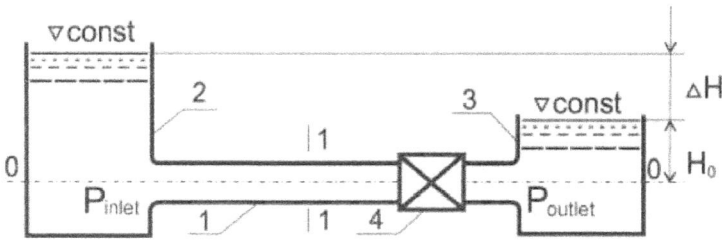

Fig. 15a. Facility used to study fluid flow in pipe

77

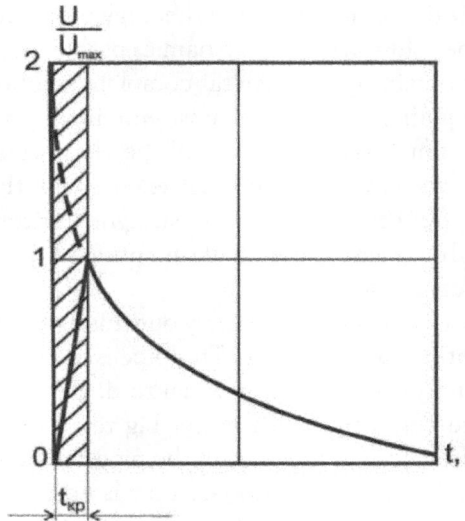

Fig. 15b. Shape of impulse of fluid flow velocity in flexible pipe in "third" flow
condition

Let us conceptually perform some experiments using the above-
mentioned facility. We shall analyze the fluid flow in the pipe at
various pressure differentials.

For this purpose, at first we shall open the gate 4 and fill the system
with water. In both reservoirs 2 and 3, which communicate through
pipe 1, the equal water levels (H0) will be established. Then we
shall close gate 4 and add some water into overhead reservoir 2.
We shall obtain Ap = PinDut~~ PoutDut pressure differential at
gate 4. Let us open the gate. The fluid in pipe 1 begins to move
under action of pressure differential Ap. Knowing the pipe diameter
and having measured the fluid velocity and water temperature, let
us calculate Reynolds number. Assume that the obtained Reynolds
number is subcriti-cal one. It means that we have a laminar flow. By
the pressure distribution along the pipe we shall find out that not
all the pipe sections have the same effect on the flow. In the initial
pipe length of about 100d, the pressure drops very rapidly, and then

pressure begins to decrease linearly lengthwise. The flow behind the initial length of the pipe is called "fully developed flow".

The first person, who has performed experimental investigations of fluid flow, was Girard. He has performed experimental investigations of water flow in brass pipes and issued results of these investigations in 1813. Girard has correctly determined the relationship between flow rate, pressure and pipe length. But he has drawn up a conclusion that the flow rate depends on the third degree of radius. Only Gagen in 1839 and Poiseuille in 1842 have found final experimental solution to the specified problem. They have shown that the flow rate depends on the fourth degree of radius. Poiseuille has carried out a number of experiments to clear up whether his law (7) is applicable to the sanguimotion in blood vessels or not.

Having carried out these experiments, Poiseuille came to be sure that the blood flowing in vessels does not obey the above-mentioned law[78] [22]. In 1883, Reynolds has showed experimentally that linear relationship between flow rate and pressure differential is lost as soon as Reynolds number reaches the critical value that is equal to 2000 for water. In this case, the flow becomes turbulent, while flow rate becomes proportional to square root of pressure differential. In this connection we would like to note that till that time no one has managed to describe the turbulent flow rigorously in terms of mathematics[79],[80][38,39].

To be more precise, no one has still managed to develop such a mathematical model that would explain all the phenomena, which are observed during experimental investigation of fluid flow in pipe within the wide range of pressure differentials. The reason for this is that there are some insuperable mathematical problems which occur when solving general equations of viscous fluid flow (Navier-Stokes equations). These mathematical problems were mentioned earlier.

78   Ref. to 55. P. 38.
79   Feinmann R., Leuton R., Sands M. Feinmann's Lectures in Physics. M., „Mir", 1977 Vol.1 P.70.
80   Turbulency. Principles & Application. Under the editorship of W.Frost, T.Moulden. M., "Mir", 1960 P.6

The author has analyzed the derivation of Navier-Stokes equations using postula-tional method. For this purpose, he represented prerequisites, which have been taken as a ground for these equations, as a system of axioms. It has been established that the above-mentioned system of axioms is inconsistent and dependent one. It had to be corrected in order to eliminate its inconsistency and dependency. Based on corrected system of axioms, the mathematical model of real fluid flow in pipe was developed. This model described the following: unsteady flow and its transformation into the steady one at subcriti-cal Reynolds numbers, the mechanism of unsteady flow stability loss at critical and supercritical Reynolds numbers, and turbulent flow. The obtained theoretical results agree completely with the experimental ones. The turbulent flow has been described in precise mathematical terms. This phenomenon was known from experiments (although, a lot of new facts have been discovered about this flow thanks to this solution), but has not been described till now. Still, the more interesting and important fact is that this solution explained how the head loss due to friction and flow pattern change in the pipe cross section from the moment of fluid motion initiation till the moment of unsteady flow transformation into the steady flow.

After instantaneous opening of gate 4 (Fig. 15a), the fluid overcomes the inertial forces and moves in the pipe with speed, which this fluid could attain under the preset pressure differential provided the fluid is ideal (Fig. 16a):

$$U_{max} = \sqrt{\frac{2\Delta p}{\rho}}$$

where
U_{max} - is velocity of fluid flow in the max pipe;
Δp - is pressure differential at pipe ends;
r - is fluid density.

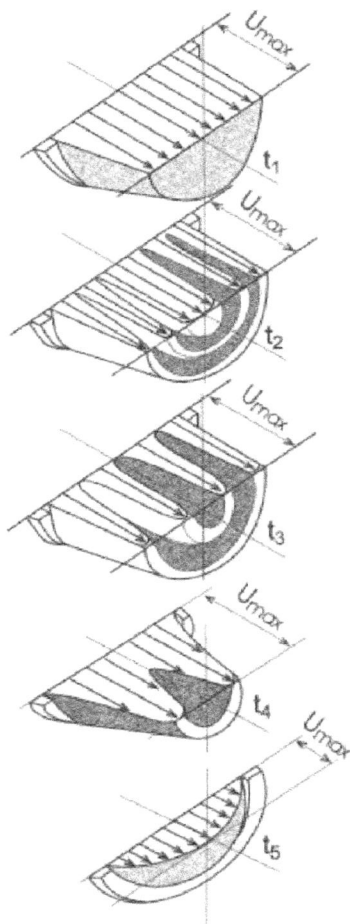

Fig.16. Diagrams of fluid flow velocities in the same flow area of the pipe at different acceleration time ($t_1 < t_2 < t_3 < t_4 < t_5$).
Here, t_5 is the time for unsteady fluid flow transformation into steady fluid.

Only very thin fluid layer that had adhered to the inner surface of the pipe remains stationary. Further, the thickness of this layer will rapidly increase due to friction. Moreover, the process of fluid flow deceleration due to friction has the wavelike properties and is carried out as follows: as soon as the fluid begins to move in the pipe, the inner surface of this pipe causes the generation of expanding packet of concentric pressure waves (Fig. 16b), or friction waves (to be more precise). In friction waves, the longitudinal velocity of fluid flow varies according to the sinusoidal law. Therefore, the friction waves separate fluid flow in the pipe into alternating concentric velocity and decelerated layers. The fluid flow velocity in these layers changes within the range of zero up to Umax. According to Bernoulli law, the static pressure in the velocity layers is less than static pressure in the decelerated layers by impact head value. The leading edge of the last wave from the above-mentioned packet of concentric pressure waves moves toward the pipe axis at the velocity of sound. At that, the radius of this packet of waves is decreasing. After the wave leading edge has reached the pipe axis, the radius becomes equal to zero and the wave is degenerated. According to the same law, this wave is followed by another wave. It is a continuous process, but the velocity of each afterwave is less than velocity of previous wave. The process of friction wave packet degeneracy on pipe axis continues until boundary ring layer transforms into axial blood stricle. Further expansion of the boundary layer will be accompanied by decrement of maximum velocity along pipe axis until the flow becomes steady, i. e. Poiseuille flow.

In this case, maximum velocity of fluid flow along pipe axis becomes several exponents less than the maximum velocity of the velocity layers in the incipient unsteady flow. Fig. 16 schematically presents the described evolution of the diagram of axial components of fluid flow velocities in the flow cross section 1-1 of the pipe (Fig. 15a) since the initiation till reaching steady (Poiseuille flow) conditions. It takes less than one second for unsteady f low tra n sfor m ation i nt o st e ady one. Moreover, the higher is the pressure differential at pipe ends the shorter is the process of transformation. The special feature of this flow is that the average velocity of the flow is one half of maximum velocity, the same as in case of Poiseuille flow. But taking into consideration the fact that maximum fluid flow rate in this case is equal to ideal flow rate (i. e. flow rate at zero friction), hydraulic efficiency of the pipe in case of this flow will be equal to 25%, while eficiency is reduced hundred times when transforming into Poiseuille flow (it is reduced thousand times when Reynolds number is very low).

From the stated above it follows that when motion is initiated in pipe, the real fluid during some fractions of a second possesses superfluidity features due to low friction losses.

This phenomenon was also detected when pipeline fluid flow initiation was researched experimentally[81],[82] [40,41].

Superfluidity condition, as it becomes clear from Fig. 16, is peculiar for high transverse gradients of velocities, and therefore of static pressure. Suspended particulate matter, e.g. erythrocytes, having got into such flow, will be popped out from the decelerated layers into velocity layers by transverse gradients of static pressure. As a result, flow structure will assume sandwich pattern. As the radii of concentric velocity layers are reduced, erythrocytes (which fill them) together with these layers flow to the pipe axis. In this case they come close and form a toral rouleau (Fig. 17).

81   Bargessyan M.G. About Friction Coefficient under Unsteady-State Movement Conditions in Pipes". Academy of Sc. of the Armenian SSR. Technical Series, M., 1971, Vol.24 No.6.
82   Coppel T.A., Liiv U.R. Experimental Investigations of Fluid Movement Initiation in Piping. Published by the Academy of Sciences of the USSR, MZhG, 1977 No.6 P.69.

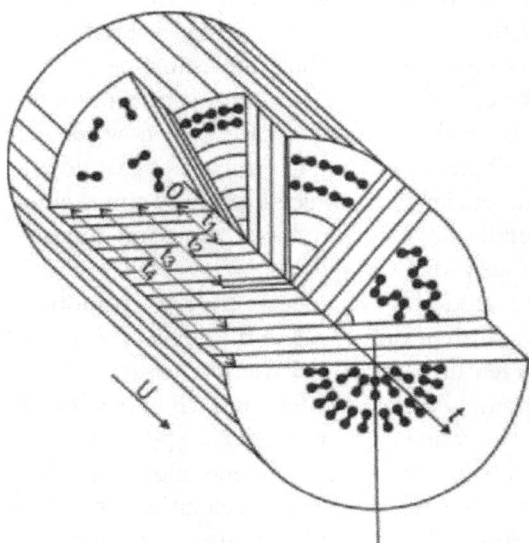

Fig. 17. Formation of concentric pattern of the flowing blood and its transformation into radial-circular one with the subsequent formation of erythrocytic rouleaux. Here t is the time of fluid flow in the pipe with constant velocity U.

After this, the forced erythrocytes motion to pipe axis is stopped. Rouleau is effected by the contracting concentric positive-pressure wave, it compresses the rouleau. As a result of compression, erythrocytes are sucked (Oros effect)[83] [25], some of them undergo erythrocytoschisis (are destroyed).

It follows from this analysis, that in order to avoid the erythrocyte hemolysis and aggregation, it is necessary to control the flow pattern in such a way that the number of concentric layers does not decrease in the course of time (as it does in the described process) and remains constant.

The solution of this problem allowed establishing the law of variation of pressure differential at pipe ends in the time domain, and

83   Ref. to 61.

the law of expansion and contraction in pipe cross section in the time domain, wherein the pattern of the flow under the superfluidity condition remains constant. With this provision, the velocity of fluid flow in the pipe must continuously decrease, as it is shown in Fig. 4. Velocity curve comprises two segments: non-steady (when $0 < t \leq t_{critical}$) and steady (when $t > t_{critical}$) segment. Flow instability is caused by indirect hydraulic impacts originated at high negative accelerations (the same as in case of rapid closing of the valve). Thus, steady flow is velocity-limited ($0 < U < U_{max}$), besides, steady flow is not durable, since it constantly decreases.

In order to avoid flow cessation, it is necessary to accelerate the flow up to the maximum velocity Umax and to repeat flow deceleration. Therefore, steady fluid flow under superfluidity conditions is of pulsed nature. Moreover, it is necessary for each impulse to expand the pipe flow area at the initial stage and then to contract it. Thus, T period of each impulse comprises three intervals:$\Delta = t_{kp}$, Δt_2 and Δt_3. During Δt_1 interval the flow accelerates up to the maximum velocity; during Δt_2 interval pipe flow area expands, during Δt_3 interval pipe flow area contracts.

So, from the stated-above it follows that steady fluid flow under superfluidity conditions differs from laminar and turbulent flows in less friction losses, steady undulating profiles of velocity and static pressure with high transverse gradients of these parameters. For this reason, this hydrodynamic condition, apart from the two familiar ones, i.e. laminar and turbulent, is designated by the authors as the "third fluidity condition".

When elaborating the third fluidity condition theory, authors established the functional links between kinematical and dynamic parameters of fluid flow and its initial parameters. Initial parameters are also designated as initial values[84] [42].

Initial parameters are as follows:

---

84   Dorodnitsyn A.A. Mathematics & Descriptive Sciences. In a Miscellany of Papers under the Title: Number & Idea. Public.No.5, edited by N.N.Moisseyev. M., "Znaniye", 1982 P.6-15.

r_0 - is a radius of pipe flow area at the beginning of expansion;
ρ - is fluid density;
a - is a sonic velocity in fluid;
g - is acceleration;
T - is hydrodynamic impulse duration;
Δt_1 - is time required to accelerate the flow up to the maximum velocity at the instant of impulse;
Δt_2 - is pipe flow area expansion interval;
Δt_3 - is pipe flow area contraction interval.

Provided the cycles follow one after the other without pauses and overlapping, then

$$T = \Delta t_1 + = \Delta t_2 + = \Delta t_3, \qquad (11)$$

Kinematical parameters of fluid flow in the pipe depend only on the specified initial parameters:

$$U_{max} = f(g, a, r_0), \qquad (12)$$

where U_{max} is maximum allowable velocity of fluid flow at the instant of impulse;

$$U = f(U_{max}, \Delta t_1, t), \qquad (13)$$

where U is fluid flow current velocity, $\Delta t_1 < t < \Delta t_3$

$$r_+ = f(r_0, \Delta t_1, t), \qquad (14)$$

where r_+ is current radius of the expanded pipe flow area, $\Delta t_1 < t < \Delta t_3$;

$$r_- = f(r_0, \Delta t_1, \Delta t_2, t), \qquad (15)$$

where r is current radius of the contracting pipe flow area, $\Delta t_1 + \Delta t_2 < t < \Delta t_3$

Dynamic parameters of fluid flow functionally depend on the same parameters as kinematical parameters, besides, they depend on fluid density[85] [43] .

Now let us consider the problem of sanguimotion in blood vessels again. We shall try to find out whether the laws of the third fluidity condition are applicable to it or not. For this purpose let us single out the main features which quite sufficiently characterize the third fluidity condition and let us identify these features in the sanguimotion conditions in vivo.

The main features of the third fluidity condition are as follows:

1. Intermittent (pulsed) fluid flow pattern.

2. Each cycle is characterized by pipe flow area expansion which is followed by contraction.

3. Flattened and M-shaped profiles of fluid flow velocities.

4. Hydrodynamic cycle falls into the following three phases: phase when the fluid accelerates up to the maximum velocity, the duration of this phase is equal to Δt_1 phase when pipe flow area expands, the duration of this phase is equal to Δt_2; phase when pipe flow area contracts, the duration of this phase is equal to Δt_3.

5. The volume concentration of the suspended particulate matter of suspension type must not exceed the volume concentration in pipe cavity filled with the velocity layers, i. e. it must not exceed 50% .

It is known from circulatory physiology that sanguimotion in blood vessels is intermittent, blood vessels in each cardio-cycle are expanded at the beginning, then they undergo contraction.

Some "in vivo" methods for measuring velocities of blood flow in arterias of animals have been developed in recent years. Using film thermal anemometry method we have detected a lot of features of velocity distribution of blood flow in the aorta of a dog. It has been ascertained that as far as aorta goes, the profile remains flattened,

85   Kisselev P.G. Hydraulics. The Fundamental Principles of Fluid Mechanics. M., «Energiya», P.295. 1980

and only the measurements carried out at aorta wall disclose that the flow is slower here[86] [3].

Measurements of velocities of blood flow in the aorta of a horse have shown that the profile of velocity at the very end of aorta during systole assumes M-shape (Fig. 16d). In thoracic aorta, high-frequency velocity fluctuations are observed, which denote the existence of flow perturbation, this phenomenon does not exist in the more distal portions of aorta[87] [3].

From the point of view of "third fluidity condition", HF fluctuations can be explained as follows. The thickness of decelerated and velocity layers in thoracic aorta is commensurable with the transducer sensitivity limit (about one millimeter). Alternating decelerated (and velocity) concentric layers execute periodic radial motion along with aorta wall, and reach the transducer which provides the velocity fluctuation readings. Aorta opening fluctuation amplitude in the distal portions of aorta is minor, the thickness of the layers is not commensurable with transducer sensitivity limit. For this reason the transducer does not read velocity fluctuations.

---

86   Ref. to 3.
87   Ref. to 3.

It is known that cardiac cycle comprises typical intervals (phases)[88],[89],[90] [44, 7, 16]. Systole falls into tension period (Tt), rapid ejection phase (Em) and slow ejection phase (Er). Diastole also falls into phases.

Blood is a kind of suspension of corpuscles (various blood cells) and specific fluid particles (chylomicrons) in plasma. The composition of blood is as follows: 50% is water, 50% are corpuscles, proteins and large number of low-molecular organic and non-organic substances. Thus, the whole set of features inherent to the "third fluidity condition" has been revealed in the sanguimotion in vivo. Besides, as it has been mentioned above, the "third fluidity condition" is distinguished by lower friction losses as compared with Poiseuille flow. So, the "third fluidity condition" satisfies the principle of optimality in biology.

It follows from the presented analysis that the laws of the "third fluidity condition" are applicable to the blood vessel sangui-motion in vivo, whereas the laws of Poi-seuille flow are not applicable to it. The concept of the blood vessel sanguimo-tion in vivo under the "third fluidity condition" has deepened the knowledge in the circulatory physiology domain. It has enlightened the specific character of cardiovascular system functioning. It has also proved that this specific character has been predetermined by the necessity to function under the most economic "third fluidity condition" of blood vessel san-guimotion. In these hydrodynamic conditions, blood flow pattern is formed by the friction against vessel internal surface so that erythrocytes are popped out by transverse gradients of static pressure from concentric decelerated layers into velocity layers where they move at maximum san-guimotion velocities. Such pattern results in the following: erythrocytes flow moves faster than plasma flow.

Deviations from the norm in the cardiovascular system functioning laws result in blood flow pattern deviation from the optimal pattern.

88   Wiggers C, Blood Circulation Dynamics. M., "IL", 1957,P.78.
89   Ref. to 11.
90   Ref. to 31.

In this case, excess of erythrocytes flow velocity over plazma velocity decreases. As a result, we observe the loss of blood circulation efficiency, which declares itself as a disease.

In addition to that, the given concept hasgiven way to the mathematization of circulatory physiology, which contributed tothe establishment of complex functionallinks between hemodynamic parameters(not available for noninvasive measurement) and duration of the cardiac cycle phases. Based on these functional links, new non-invasive methods for the determination of functional state of the portions of cardiovascular system[91] [15] can be developed.

2.3. Mathematical methods for determination of minute, stroke and phase volumes based on duration of cardiac cycle

Volume characteristics of heart, such as minute volume, stroke volume and phase volume define the pumping ability of the heart. In this connection, the information about volume of blood ejected from heart or incoming to heart within a definite time interval is of great importance for physiological, clinical and diagnostic investigations. This information is also necessary for objective assessment and monitoring of cardiovascular system functioning in normal and pathological conditions.

By now, a great number of invasive and noninvasive methods are developed for determination of minute volume and stroke volume of the heart. There is a number of methods which are based on Fick law. All methods include the stage of recording the change in concentration of various blood indicators. This stage is followed by calculation of minute volume by formulas. This principle is fundamental for direct and indirect Fick methods, indicator-dilution method (including the thermodilution and radiocardiogra-phy methods), methods of introduction of foreign gases. The oxygen method of Fick and indicator-dilution methods are widely spread now and recognized by research workers as most precise. Let us focus on them.

---

91   Patent No. 94031904 (RF). Method of Determination of the Functional Status of the Left Sections of the Heart & their Associated Large Blood Vessels. Authors: Poyedintsev G.M., Voronova O.K.

Oxygen method of Fick is based on determination of blood volume required for transportation of oxygen in amount absorbed by the lungs during one minute, it also includes the investigation of oxygen quantity in arterial blood and venous blood[92],[93],[94],[95] [21, 45, 29, 22]. Indicator-dilution methods consist in introductionof an indicator into bloodstream followedby registration of indicator concentration in blood[96],[97],[98],[99],[100] [21, 45, 29, 22, 46].

It is wise to apply these methods only under condition that the catheterization of heart cavities and great vessels shall be made[101],[102] [21, 45], but these manipulations make procedure more complicated and very unpleasant for a patient. Moreover, these methods may cause serious errors in determination of minute volume due to dramatic change in he-modynamic parameters resulted from examination procedure itself [103],[104],[105],[106][21, 22, 28, 47].

A conclusion can be made that Fick methods and dilution of indicators introduce considerable changes to hemo-dynamic characteristics which do not allow to assess the true values of minute volume and therefore to verify the data obtained through other methods. All the above-mentioned methods have two more serious drawbacks,

92   Gaiton A. The Minute Heart Volume & its Regulation. M., "Medicina», 1969, P. 26-38
93   Rashmer R., Dynamics of the Cardiovascular System. M., "Medicina", 1981. P. 80-83.
94   Cardiology Manual, under Editorship by Ye.I.Tchazov. M., "Medicina", 1982 Vol. 1 P.83-84.
95   Savitsky N.N. Biophysical Principles of Blood Circulation & Clinical Methods of Researches in Hemodynamics. L., "Medicina", 1974, P/ 205-208.
96   Ref. to 92. P. 48-74.
97   Ref. to 93. P. 33-84.
98   Ref. to 94. P. 75-79, 412-414.
99   Savitsky N.N. Biophysical Principles of Blood Circulation & Clinical Methods of Researches in Hemodynamics. L., "Medicina", 1974, P.208-215.
100   Folkov B., Nil E., Blood Circulation. M., "Medicina", 1976, P.70.
101   Ref. to 92. P. 28, 57.
102   Ref. to 93. P. 77,82,91.
103   Ref. to 92. P. 39-45, 75-78.
104   Ref. to 99. P. 208, 214.
105   Folkov B., Nil E., Blood Circulation. M., „Medicina", 1976, P.70.
106   Emmerich J.u.a. Uber den Einf luss blutiger Untersuchungsmethoden auf das Hers minutenvolumen.-Zschr. Kreislauff., 1958, Bd. 47, S. 236.

namely: first, they ignore the fact that erythrocytes average speed in vessels is higher than average speed of blood plasma[107],[108],[109],[110] [46, 36, 48, 49] that brings about the errors in determination of minute volume; secondly, they do not furnish the information about value of cardiac output per one cardiac cycle.

Some physical methods are known for determination of stroke volume and minute volume, such as sphygmography, ballistocardiography and roentgenography.

Sphygmographical method is based on analysis of arterial pulse curve and includes the calculation of stroke volume value based on empirical formulas, among which formulas of Bremzera-Ranke and Vetslera-Bogera[111],[112],[113] [21, 45, 22] are well-known. Sphygmographical and ballistocardio-graphical methods[114],[115],[116] [21, 45, 22] have a very important advantage which lies in the fact that the patient under examination is not subjected to ill-favored effects. But the shortcoming of the methods is the low level of accuracy, which is the reason for their very restricted employment in practical medicine.

Roentgenography is the most successful of all physical methods[117],[118] [21, 45]. It is distinguished by high level of accuracy, but the patient is subjected to comparatively high doze of X-radiation. Besides, the radiographic contrast substance introduced into blood bed bears a hazard condition for the patient.

---

107   Ref. to 100. P. 49.
108   Levtov V.A., Regirer S.A., Shadrina N.Kh. Blood Rheology. M., JMedicina", 1982. P.115.
109   FahraeusR.-Physial.Rev., 1929,9,231.
110   Lipowsky H.H., Kovalcheck S., Zweifach B.W. - Circulat, Res., 1978, 43, N. 5, 738.
111   Ref. to 92. P. 91-103.
112   Ref. to 93. P. 85-86.
113   Ref. to 99. P. 215-225.
114   Ref. to 92. P. 101-108.
115   Ref. to 93. P. 84-85
116   Ref. to 99. P. 225-226.
117   Ref. to 92. P. 108-113.
118   Ref. to 93. P. 69-71.

In recent years, the ultrasonic investigation and rheography are more and more in the focus of attention of specialists for measuring the stroke volume and minute volume.

Ultrasonic methods of investigation[119],[120],[121] [29, 50, 51] are based on ultrasonic measuring of linear dimension of heart ventrical followed by calculation of stroke volume using empirical formulas, the most exact of which is the formula introduced by Teichholz and co-authors[122],[123] [29, 51]. With this, it is assumed that the ventricular intracavity has a shape of an oblong ellipsoid. Further development of ultrasonic technology and invention of B-and M-scanning technique extended the boundaries of method application and improved accuracy of measurements. Still, some major mistakes are probable in using the ultrasonic method for measuring of stroke volume due to a) some simplifications in determination of ventricular intracavity shape[124] [50]; b) peculiarities of the heart spacial dynamics in systolic and diastolic phases[125] [52]; c) errors in determination of the heart linear size.

Rheography method[126],[127],[128],[129] [29, 53, 54, 55] is extensively used in practical medicine. Though it has positive advantages, some sig-

119   Ref. to 94. P. 130.
120   Zaretsky V.V., Bobkov V.V., Olbinskaya L.I. Clinical Echocardiography. Atlas. M., „Medicina", 1979. P. 232-231.
121   Mukharlyamov N.M., Belenkov Yu.N. Ultrasound Diagnostics in Cardiology. M., „Medicina", 1981. P. 59-62.
122   Ref. to 94. P. 130.
123   Ref. to 121. P. 60.
124   Ref. to 120. P. 229.
125   Karpman V.L., Sinyakov V.S. Three-DimensionalDynamics of the Left Ventricle and Phase Structure ofthe Heart Cycle // Physiol. Journal. USSR.1965.13.n.51..N°7.P.832-837.
126   Ref. to 94. P. 51-54.
127   Kubicek W.G., Petterson R.P., Witson D.A. et al. Development and Evaluation of an Impedance Cardiac Output System. Aerospace Med. 1966.V.37 No.12 P.1208-1212.
128   Kubicek W.G., Petterson R.P., Witson D.A. Impedance Cardiography as a Noninvasive Method of Monitoring Cardiac Function and Other Parameters of the Cardiovascular System. Ann.N.V.Acad.Sci.1970 V.170P.724-732.
129   Tischenko M.I., Smirnov A.D., Danilov L.N. et al. Characteristics and Clinical Application of Integral Rheography - a New Method in Measuring Stroke Volume. "Cardiologiya", 1973.

nificant errors are probable in determination of stroke volume of the heart.

All the methods (except for echocardiography and roentgenography), the brief description of which is provided in this section, do not facilitate the determination of phase volume values of heart, thus decreasing their informative and diagnostic value. With a knowledge of minute and stroke volumes only it is impossible to assess adequately the function of heart, since even essential changes in functional state of the heart may not be accompanied with changes in values of these parameters[130] [45]. The most objective assessment of pumping ability of the heart can be made through complex analysis of minute, stroke and phase volumes of the blood circulation.

In terms of the above-mentioned information, we can make a conclusion that development of new means for determination of cardiovascular system volumetric parameters is still a matter of vital necessity.

Entirely new approach to this issue was applied based on concept of third condition of blood flow in vessels. The detailed description of the concept is given in Section 2.2, we use the same terminology and parameters identification as adopted by G. Poyedintsev in his paper.

Let us consider the algorithm in more detail.

Let us assume that:

1. Maximum fluid flow rate in elastic tube within the impulse is determined by the following equation:

$$U_{max} = f(g, a, r_0).$$

---

130   Ref. to 93. P. 80.

2. Momentary fluid flow rate in elastic tube changes within the impulse by the following law:

$$U = f(U_{max}, \Delta t_1, t),$$

where $\Delta t_1 < t \leq \Delta t_3$.

3. Radius of elastic tube clearance at the instant of tube expansion is changed by the following law:

$$r_- = f(r_0, \Delta t_1, t),$$

where $\Delta t_1 < t \leq \Delta t_2$.

4. Radius of elastic tube clearance at the instant of tube contraction is changed by the following law:

$$r_- = f(r_0, \Delta t_1, t),$$

where $\Delta t_1 < t \leq \Delta t_2$.

Proceeding from this, the formulas were deduced which determine the following:

1. Quantity of fluid flowing through cross section of elastic tube within the period of its expansion is determined as follows:

$$Q_1 = \int_{\Delta t_1}^{\Delta t_1 + \Delta t_2} \pi \cdot r_+^2 \cdot U \cdot dt = \quad (16)$$

$$= \pi \cdot r_0^2 \cdot f(g, a, \Delta t_1, \Delta t_2) ,$$

2. Quantity of fluid flowing through cross section of elastic tube within the period of its contraction is determined as follows:

$$Q_1 = \int_{\Delta t_1 + \Delta t_2}^{\Delta t_1 + \Delta t_2 + \Delta t_3} \pi \cdot r_-^2 \cdot U \cdot dt = \qquad (17)$$

$$= \pi \cdot r_0^2 \cdot f(g, a, \Delta t_1, \Delta t_2, \Delta t_3),$$

When fluid flows in elastic tube, fluid inflow during tube expansion exceeds fluid outflow by volume equal to ΔQ, while the outflow during tube contraction exceeds the inflow by the same volume equal to ΔQ. Volumes Q_1 and Q_2 define the outflow.

We obttain:

1. Quantity of fluid going into elastic tube within the period of tube expansion:

$$Q_3 = Q_1 + \Delta Q \quad (18)$$

2. Quantity of fluid going into elastic tube within the period of tube contraction:

$$Q_4 = Q_2 - \Delta Q \quad (19)$$

So, we have obtained the mathematical model which enables deduction of the formulas for determination of systolic and di-astolic volumetric parameters of pumping ability of heart.
In conclusion, let us remark here that:
– in the first place, elastic tube functioning law, which is dictated by necessity to effect the motion of fluid in tube according to third con-

dition law, imposes the special requirements for supply system, this is a factor which strictly determines the volume values Q_3 and Q_4;
– in the second place, all numerical coefficients which will be included in formulas for determination of volumetric parameters were obtained not empirically but as a result of parameters values substitution ($g = 9.81$ m/s2 and $a = 1585$ m/s) and upon accomplishing of various mathematical operations.

Let us proceed to description of derivation of formulas to be used for determination of systolic volume characteristics.

Let us substitute the following parameters for the terms of equations (16) to (19):

$$\Delta t = AC + IC;$$

$$\Delta t = Em; \Delta t3 = Er; S = \pi r_0^2,$$

where

AC - is duration of phase of asynchronous contraction (Q - R);
IC - is duration of phase of isometric contraction (R - S);
Em - is duration of phase of rapid ejection (L - j);
Er - is duration of phase of slow ejection (j - T);
S - is cross-sectional area of ascending aorta. This area is determined by nomogram or by other means.

We obtain equation for Q_3 and Q_4, which in this case are equal to blood volume PV1 and PV2 respectively, which arrive to ascending aorta (expelled from ventricular) in phases of rapid and slow ejection.

Then we define derived values.

We omit conclusions and go to end formulas:

$$PV_1 = S \cdot (AC + IC)^2 \cdot f_1(\alpha) \cdot [f^2(\alpha) +$$

$$+ f_3(\alpha, \beta, \gamma, \delta)], \; (ml);$$

$$PV_2 = S \cdot (AC + IC)^2 \cdot f_1(\alpha) \cdot$$

$$\cdot f_4(\alpha, \beta, \gamma, \delta), \; (ml),$$

where $f_1(\alpha) = \dfrac{22072.5[(5\alpha - 2)^3 - 27]}{(5\alpha - 2)^5 - 243}$;

$$f_2(\alpha) = \dfrac{\alpha^5 - 1}{2};$$

$$f_3(\alpha, \beta, \gamma, \delta) = \dfrac{1}{8}[\dfrac{10}{3}(4\alpha^2 - \delta^2)(\beta^3 - \alpha^3) +$$

$$+ 5\chi\delta(\beta^4 - \alpha^4) - 2\chi^2(\beta^5 - \alpha^5)];$$

$$f_4(\alpha, \beta, \gamma, \delta) = \dfrac{1}{8}[5(\delta^2 - \dfrac{8}{3}\alpha^2)(\beta^3 - \alpha^3) +$$

$$+ 7,5\chi\delta(\beta^4 - \alpha^4) + 3\chi^2(\beta^5 - \alpha^5)];$$

$$\alpha = (1 + \dfrac{Em}{AC + IC})^{0.2};$$

$$\beta = (1 + \dfrac{Em + Er}{AC + IC})^{0.2};$$

$$\chi = \dfrac{2(\alpha - 1)}{\beta - \alpha};$$

$$\delta = \alpha(2 + \chi).$$

84

Stroke volume (SV) of heart is equal to:

$$SV = PV_1 + PV_2 =$$

$$= S \cdot (AC + IC)^2 \cdot f_1(\alpha) \cdot [f_2(\alpha) +$$

$$+ f_3(\alpha, \beta, \gamma, \delta) + f_4(\alpha, \beta, \gamma, \delta)], \text{ (ml)}$$

or:

$$SV = S \cdot q_s, \text{ (ml)},$$

where q_s is impact density of blood flow, i.e. blood volume falling on 1 cm^2 of ascending aorta lumen area per one cycle.

Minute volume (MV) is equal to:

$$MV = SV \cdot PR = S \cdot q_s \cdot PR = S \cdot q_M,$$

where q_M is blood flow density per minute, i.e. blood volume falling on 1 cm^2 of ascending aorta lumen area per minute; PR is pulse rate. Parameters qs and q_M are self-dependent values. They were determined and analyzed during investigation process.

Phase volumes PV_1 and PV_2 can be directly determined in percentage of SV, which is taken as 100%:

$$PV_1 = \frac{100[f_2(\alpha) + f_3(\alpha,\beta,\chi,\delta)]}{f_2(\alpha) + f_3(\alpha,\beta,\chi,\delta) + f_4(\alpha,\beta,\chi,\delta)} \text{ (\%)};$$

$$PV_2 = 100 - PV_1 \text{ (\%)}.$$

When determining values of MV and SV in this way, the volume of blood arriving into coronary vessels is not taken into account. As is known, this volume constitutes about 5% of total blood volume ejected by heart[131] [56].

131   Lightfoot E. Phenomena of Transfer in Live Systems. M., "Mir", 1977. P.21.

Now we can briefly describe derivation of formulas for determination of diastolic volume parameters.

Diastole is considered as two self-dependent hydrodynamic processes. The first one consists in rapid and slow filling of ventricular (similar to rapid and slow ejection in systole). In this case, the following parameter values are inserted into equations (16) to (19):

$$\Delta t_1 = P + IR;$$

$$\Delta t_2 = Fr;$$

$$\Delta t_3 = Dy,$$

where

P is duration of protodiastole phase $(T - U_H)$;
IR is duration of isometric relaxation phase $(U_H\text{-}AU)$;
Fr is duration of rapid filling phase $(AU \cdot U)$;
Dy is duration of slow filling phase or dia stasis $(U \cdot P_H)$

The second process consists in atrium systole (is considered similar to rapid ejection in systole). In this case, the following parameter values are inserted into equations (16) to (19):

$$\Delta t_1 = t_0,$$

$$\Delta t_2 = Sa,$$

$$\Delta t_3 = 0,$$

where

t_0 is a time interval determined theoretically based on specified correlations;

Sa is duration of atrium systole.

After some transformations had been accomplished, the formulas were obtained for determination of specific phase volume values in diastole, namely blood volumes PV_3, PV_4, PV_5, which arrived in cardiac ventricle in phases of rapid and slow filling and in phase of atrium systole. Phase volume values are represented in percentage of entire filling volume which is taken as 100% :

$$PV_3 =$$
$$= 100 \cdot (P + IR)^2 \cdot f_1(\alpha)[f_2(\alpha) + f_3(\alpha,\beta,\chi,\delta)] /$$
$$/\{(P + IR)^2 \cdot f_1(\alpha) \cdot [f_2(\alpha) + f_4(\alpha,\beta,\chi,\delta)] +$$
$$+ t_0 \cdot W \cdot f_6(\varepsilon)\} \, (\%);$$

$$PV_4 = 100 \cdot (P + IR)^2 \cdot f_1(\alpha) \cdot f_5(\alpha,\beta,\chi,\delta) /$$
$$/\{(P + IR)^2 \cdot f_1(\alpha) \cdot [f_2(\alpha) + f_4(\alpha,\beta,\chi,\delta)] +$$
$$+ t_0 \cdot W \cdot f_6(\varepsilon)\} \, (\%);$$

$$PV_5 = 100 - PV_3 - PV_4 \, (\%),$$

where $f_1(\alpha) = \dfrac{220.725[(5\alpha - 2)^3 - 27)]}{(5\alpha - 2)^5 - 243}$;

$$f_2(\alpha) = \frac{\alpha^5 - t}{2};$$

$$f_3(\alpha,\beta,\chi,\delta) = \frac{1}{8}[\frac{10}{3}(4\alpha^2 - \delta^2)(\beta^3 - \alpha^3) +$$
$$+ 5\chi\delta(\beta^4 - \alpha^4) - 2\chi^2(\beta^5 - \alpha^5)];$$

$$f_4(\alpha,\beta,\chi,\delta) = \frac{1}{8}[\frac{5}{3}\delta^2(\beta^3 - \alpha^3) -$$

$$-\frac{5}{2}\chi\delta(\beta^4 - \alpha^4) + \chi^2(\beta^5 - \alpha^5)];$$

$$f_5(\alpha,\beta,\chi,\delta) = \frac{1}{8}[5(\delta^2 - \frac{8}{3}\alpha^2)(\beta^3 - \alpha^3) -$$

$$-7,5\chi\delta(\beta^4 - \alpha^4) + 3\chi^2(\beta^5 - \alpha^5)];$$

$$\alpha = (1 + \frac{Fr}{P + IR})^{0.2};$$

$$\beta = (1 + \frac{Fr + Dy}{P + IR})^{0.2};$$

$$\chi = \frac{2(\alpha - 1)}{\beta - \alpha};$$

$$\delta = \alpha(2 + \chi).$$

As it has been mentioned already, the value t_0 is determined theoretically. First of all, the value of parameter W is determined. This value must comply with the following equality:

$$\frac{B\{[5(1 + AW^2)^{0.2} - 2]^3 - 27\}}{W^3\{[5(1 + AW^2)^{0.2} - 2]^3 - 243\}} = 1,$$

where $A = \dfrac{Sa}{(P + IR)^3 \cdot f_1^2(\alpha) \cdot [\alpha + \frac{\chi}{2}(\alpha - \beta)]};$

$$B = \frac{220.725 \cdot Sa}{A}.$$

Then the other parameters are determined:

$$t_0 = \frac{Sa}{A \cdot W^2};$$

$$\varepsilon = (1 + \frac{Sa}{t_0})^{0,2};$$

$$f_6(\varepsilon) = \frac{\varepsilon^5 - 1}{2}.$$

On the ground of presented in this section mathematical relationships between volumetric characteristics of heart pumping ability and cardiac cycle phases duration, a number of new noninvasive methods for determination of cardio-vascular system state has been developed. These methods have been awarded certificate of authorship[132] [15].

For the purpose of practical application of algorithms for hemodynamic parameters calculation in the medical tools, the following designations have been introduced that also facilitate the diagnostics.

The volumes of blood entering the heart during the phases of rapid and slow filling have been summed into one group and designated as PV_1 (ml). This volume corresponds to early diastole. From our point of view, this term (designation) meets the practical requirements of diagnostics to the best advantage.

The volumes of blood corresponding to the other phases have been designated in the similar way:

---

132   Patent No.94031904 (RF). Method of Determination of the Functional Status of the Left Sections of the Heart & their Associated Large Blood Vessels. Authors: Poyedintsev G.M., Voronova O.K.

PV$_2$ (ml) for the volume of blood entering the ventricle during atrial systole

PV$_3$ (ml) for the volume of blood ejected by ventricle during the rapid ejection phase

PV$_4$ (ml) for the volume of blood ejected by ventricle during the slow ejection phase

PV$_5$ (ml) for the volume of blood ejected by ascending aorta when it is functioning as a peristaltic pump.

The sums PV$_1$ + PV$_2$ and PV$_3$ + PV$_4$ can be used to check the calculations. They must be equal.

2.4. Theorem of identification of cardiac cycle phase transition parameters

Hemodynamic parameters of cardiovascular system can be measured to high precision and reliability level only after the problems related with identification of cardiac cycle phases criteria are solved. For this purpose, let us try to amplify the concept proposed by G. Po-edintsev and O. Voronova with some theoretical considerations, which shall logically link up the phase structure of ECG (electrical signals) with mechanical response of vessels that is characterized by the rheogram.

In general, the cardiovascular system performance can be presented as a piston movement in a rigid pipe. If the piston executes sinusoidal oscillation, the fluid in the pipe will oscillate in the same way. In this case, the phases of direct flow and reverse flow will alternate. The shape of the rheogram reflects the return oscillation that can be observed in the real blood flow. The diastolic portion of the rheogram demonstrates this phenomenon in the most obvious way. According to the theory proposed by G. Poedintsev and O. Voronova, this periodic return oscillation is aimed at forming the concentric rings of blood elements, and this very oscillation enables high fluidity of blood.

The piston movement is a harmonic oscillation. It is easy to register the instant of the harmonic oscillation transition from the direct motion into reverse motion and vice versa. This very instant is a criterion of the motion phase boundary that can be also observed during the heart work. This instant can be fixed very precisely after the piston movement curve differentiation. In the piston movement curve, the initial reversal point will correspond to the extremum point of the derivative curve. In this very point the function shall reverse the sign.

The process of the piston movement during one sinusoidal oscillation cycle is represented in Fig. 18. The first-order derivative is provided in the same figure. For the purpose of clarity, let us denote the piston movement curve by sin(x), and the derivative of this curve by dsin(x)/dt. We are interested in obtaining the representation of that stage of the above-mentioned process when the piston changes the direction of its motion to the reverse one. This instant corresponds to the time value t_2.

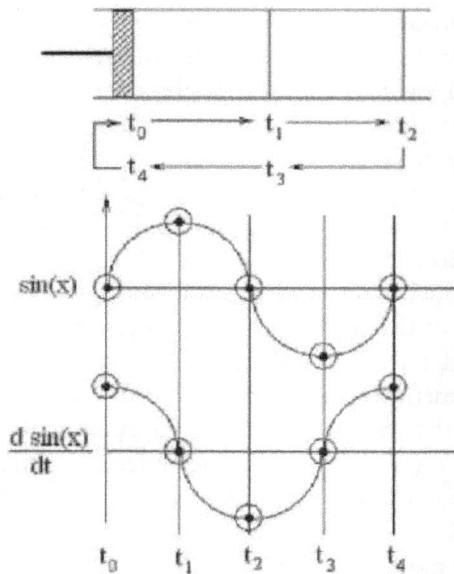

Fig. 18. Sinusoidal movement of piston and the curve of piston movement with synchronous representation of the first-order derivate during various phases

In this case, the extremum of the deri-vate lies in the minimum, and the function changes the sign to the opposite one. This indicates the piston reversion. During this period, the motion was comprised of two phases: during the first phase, the motion velocity increment was positive (as is seen from the positive part of the derivate curve); and during the second phase, the positive increment was reduced to zero. The instant when deceleration is initiated corresponds to the curve inflection point tr At this instant, the derivate value changes the sign from positive to negative.

The second period is a mirror presentation of the period described above. The only difference is that during the first phase of this period, the derivative rises and becomes positive at the instant t3. This indicates the development of the motion process. But further, the motion increment decreases, and at the instant of time t4 the piston stops for a while to begin the motion in reverse direction. At this instant, the derivative will lie in the positive extremum. This process can be repeated many times.

The fact of the compliance of the $\sin(x)$ curve inflection points with the extrema of the first-order derivative is a mathematical low[133] [57].

Biomedical signals are compound signals that comprise many harmonic sinusoidal signals of various frequencies. But in order to identify the curve inflection points and extrema with high accuracy, these signals can be analyzed using derivatives.

This fact is proved in the paper devoted to the investigation of the interference of arterial pressure waves in the occlusal blood flow[134] [11].

While transferring the above-mentioned investigation process to the process of searching for criteria applicable to cardiac cycle phases recording, the strategy for the further practical proof of the proposed theory shall be taken into account. In this case, when the theory of the "third high fluidity condition" is used to obtain the final calculated volumetric hemodynamic values, it is necessary to compare these values with the values obtained using the earlier known methods, for example ultrasonic echocardiography is the most suitable method. In this case, the values obtained using this method can be assumed as reference values.

---

133   Bronshtein I.N., Semendyaev K.A. Reference Book in Mathematics. M., "OGIZ", 1948 P.309-328.
134   Rudenko M.Yu., Alexeyev V.B., Matsyuk S.A. Biophysical Phenomena in Blood Circulation in Indirect Measuring Arterial Pressures & Evaluation of the Relevant Instrumentation // „Medtechnica". 1986 No.5 P.26-35.

ECG investigation based on the derivatives is carried out using computer programs which enable to obtain both quantitative and qualitative data. Fig. 19 provides the real ECG curve and ECG derivative. It can be seen that the derivative is repeated in each cardiac cycle. This indicates the availability of information about specific phase structure of the heart work. When analyzing these specific features, the following fact shall be taken into account: all cyclic processes which take place in human organism are similar to other natural processes. In particular, there are no sharp changes in the biorhytmical processes. The processes vary smoothly, in cycles, like sea waves in a big ocean. To be more precise, the bursts and collapses in the natural processes energy can be precisely analyzed using the mathematical differentiation.

The alternation of cardiac cycle phases is also a harmonic process. Each phase has its own progression and attenuation process. **Respectively, the boundaries of phases transition shall correspond to the ex-trema of the first-order derivative.** This point is of great importance and should be assumed as the axiomatic ground of the further theory, which needs to be proved in practice.

If we accept this fact a priori, the cardiac cycle phases presented in Fig. 19 shall correspond to the derivative extrema. Thus, the following points are clearly heightened in the first-order derivative curve: P, Q, S, T. These points shall be assumed as criteria of the beginning and end of the corresponding phases. Point S that marks the end of the QRS complex is an important criterion. At this instant, the process of the semilunar aortic valves opening is initiated and blood ejection from left ventricle to aorta begins. This process requires high energy cost. Therefore, this process is most subject to disturbance in case of cardiovascular system pathologies.

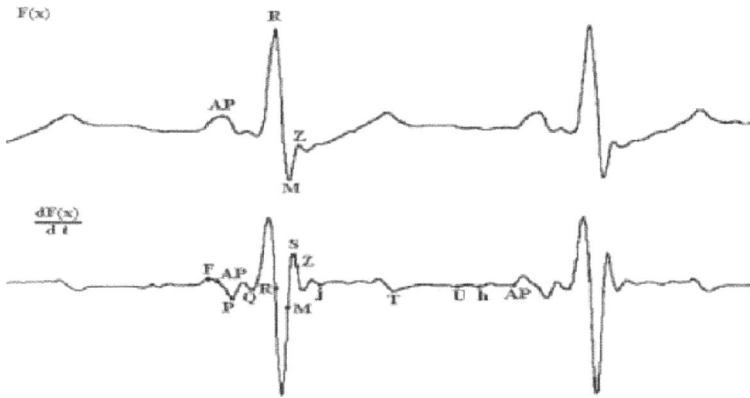

Fig. 19. Curve and first-order derivative of ECG. All the signals are standardized and do not have distorted information

Earlier on, point S was defined as intersection of the ECG ascending branch (in the specified phase) and the isoline drawn from point Q which indicates the beginning of the systolic time interval (Fig. 20). It is a disputable matter, still more disputable if we take into account the fact that point Q itself can not always be evident in ECG. The location of point Q can be approximately determined in the third (III) or fourth (IV) lead of ECG. But it is a big problem to determine the location of this point in other leads. Therefore, the generally accepted criterion of point S registration was accepted rather tentatively than based on proofs. Due to the low accuracy of measurement, such an approach has not stimulated the development of phase analysis in practice.

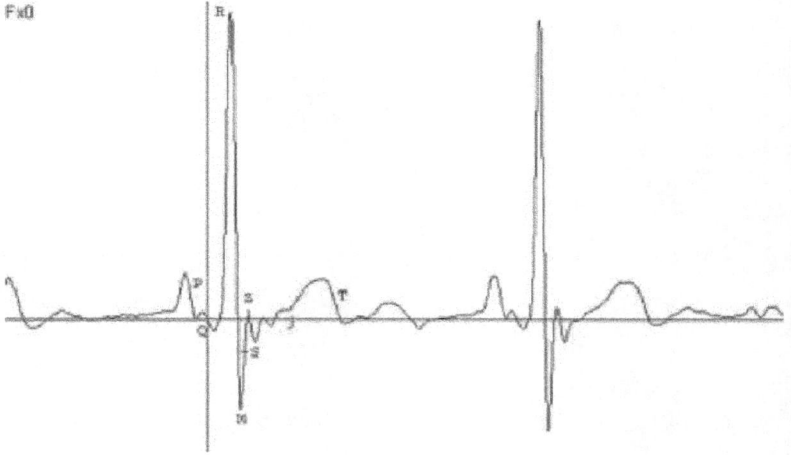

Fig. 20. Determination of point S on ECG trace through ECG ascending branch intersection with the isoline drawn from point Q

Point S shall correspond to the instant when semilunar aortic valves open and arterial pressure begins to rise in aorta. To prove this fact, the ECG, its first-order derivative and the ascending aorta rheogram were recorded simultaneously. The curves obtained as result of this record are provided in Fig. 21.

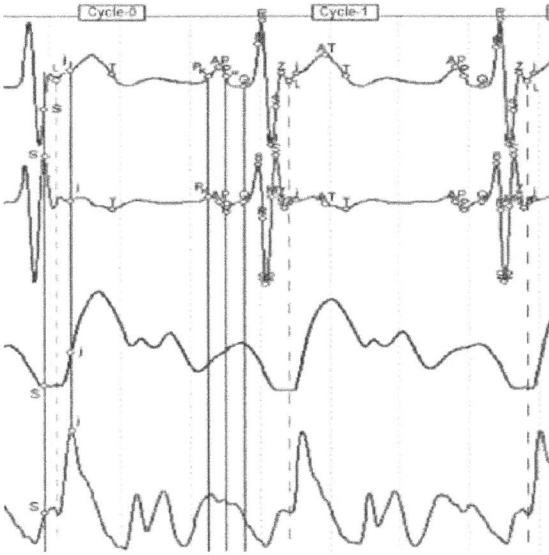

Fig. 21. Synchronous recording of ECG, its first-order derivative and the rheogram of the ascending aorta with derivate

It is clearly seen in this Figure, that the extremum of ECG derivative in point S corresponds to the instant of the aortic pressure rise. This proves the adequacy of the theoretical concepts. To prove this fact, more than 500 patients were investigated and results were the same in 100% of cases.

The adequate choice of point S registration criterion allows to discuss other ECG points. First of all, let us consider point Q. This point corresponds to the instant of atrioventricular valves closing and to the instant of the beginning of the myocardium contraction process that leads to build-up of pressure in the cardiac ventricles and ends in point S. Point Q also corresponds to the extremum, but in this case, the extremum is negative.

Point P is localized in the same way, it also corresponds to the negative extremum.

97

It should be noted that information criteria of the beginning and end of cardiac cycle phases are located in ECG inflection points, but not in the ECG extrema. This corresponds to the above-mentioned concept of the smooth transition of the energy processes from one phase to another.

The critical point is point T. This point corresponds to the instant when the semi-lunar valves begin to close and the aortic pressure begins to drop. Respectively, the negative extremum is registered on the derivative. This fact is also proved by the rheogram on which the drop in maximum pressure can be observed.

ECG analysis using derivative was checked through investigation of 1000 patients by clinical tests. Only one lead was used during the tests (i. e. lead developed within the scope of this work). The specified criteria were registered on derivative in 100% of cases, both in norm and pathology. The clear and legible registration of the criteria enable not only to ensure high reliability of the cardiac cycle phases recording but also to computerize fully the process of phases duration calculation and process of the instantaneous calculation of 15 quantitative hemodynamic characteristics based on equations proposed by G. Poed-intsev and O. Voronova.

2.5. ECG phases representing the myocardium electric biopotential correlation with heart geometric shape variation

High-accuracy research of heart work mechanism enabled to obtain information, which proves the existence of 10 phases in changing the heart shape within a single cardiac cycle. Fig. 22 shows the phases of ECG which characterize the phase mechanism of heart and vessels work.

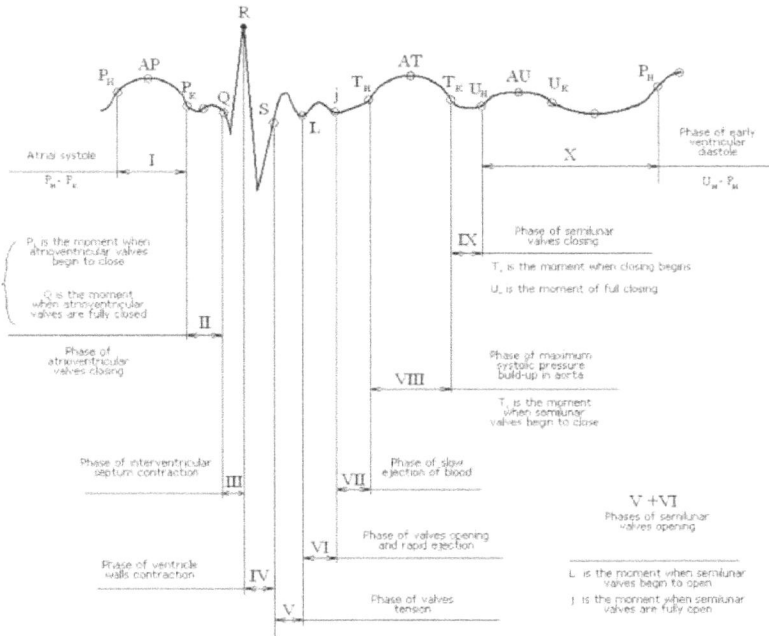

Fig. 22. Phase structure of ECG trace

For thorough understanding of diagnostic capabilities of the presented information, let us consider the relationship between cardiac cycle phases shown on ECG trace and the associated physical parts of heart, which are active in the given phase.

2.5.1. Heart geometric shape versus cardiac cycle phases

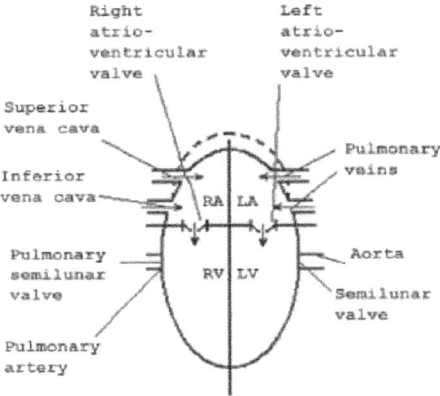

Fig. 23. Blood volume increasing in ventricles in phase $P_H - P_к$

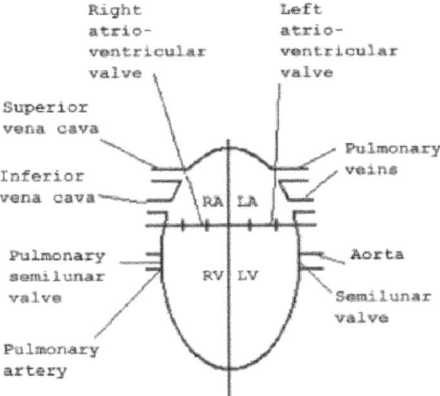

Fig. 24. Interventricular septum contraction in Q–R phase

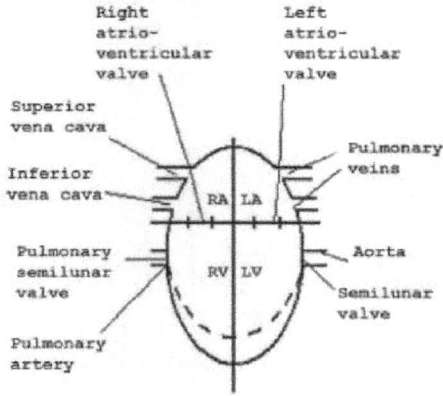

Fig. 25. Atrioventricular valves closing in P–Q phase

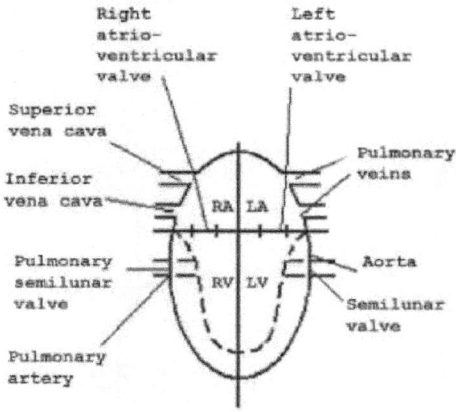

Fig. 26. Ventricles walls contraction in phase R–S

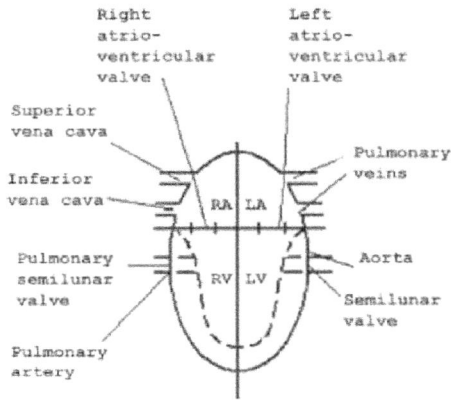

Fig. 27. Semilunar valves tension in S–L phase

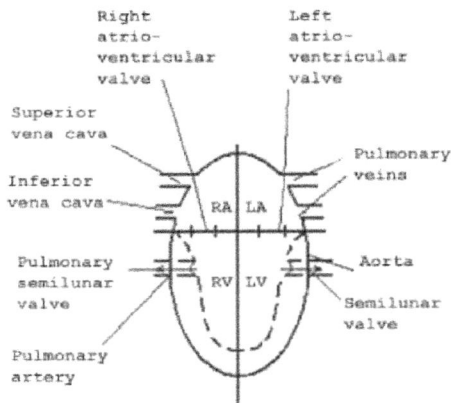

Fig. 28. Semilunar valves opening and rapid ejection in L–j phase

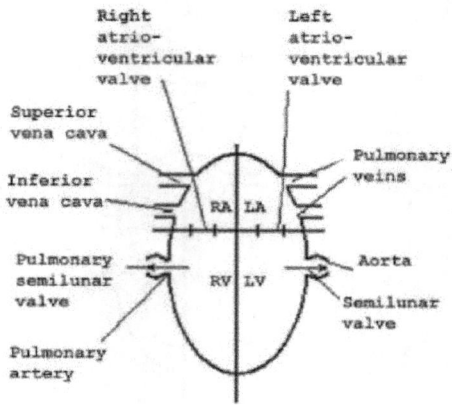

Fig. 29. Slow ejection in j–T_H phase

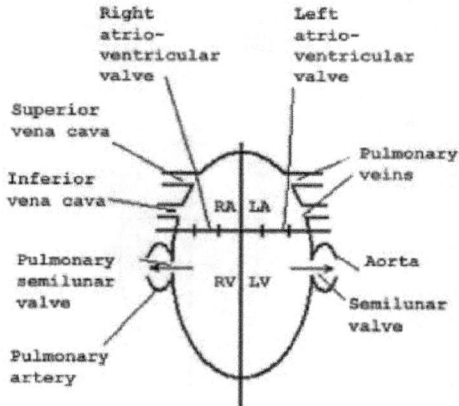

Fig. 30. Build up of maximum systolic pressure in aorta in T_H–T_k phase

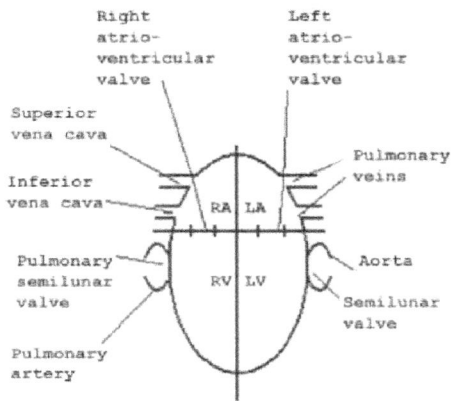

Fig. 31. Semilunar valves closing in T –U phase

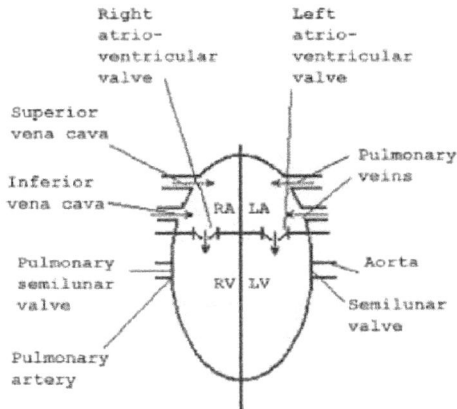

Fig. 32. Atrioventricular valves opening at instant U_H and ventricles fillingin early diastole phase $U_H – P_H$

2.6. Delimitation of phases using the curve of first-order derivative for ECG when conducting mathematical differentiation

2.6.1. Graphic differentiation of functions in classical mathematics

In Section 2.4 we considered the theory of cardiac cycle phase transitions and proposed the principle of revealing their criteria using graphic mathematical differentiation. Taking into account that mathematical differentiation is a very complicated task for the doctors, we need to consider it in detail.

Graphic differentiation is very popular in mathematics. With the aid of differentiation we can investigate different implicit functions which are difficult to describe by mathematical equations. The essence of graphic differentiation is the determination of singular points of the function, namely its inflection points and local ex-trema. As a rule, double differentiation is applied. Having obtained two curves of derivatives of the function, it becomes possible with high probability to define regular processes which generate the curve of the function. This is one of the essential research means which is used in experimental scientific works.

Let us consider graphic differentiation principle with an example of elementary sinusoidal harmonic signal (item 2.4).

Figure 33 shows the curves of function sinX and its derivative cosX. The curve of function cosX is the first-order derivative for function sinX.

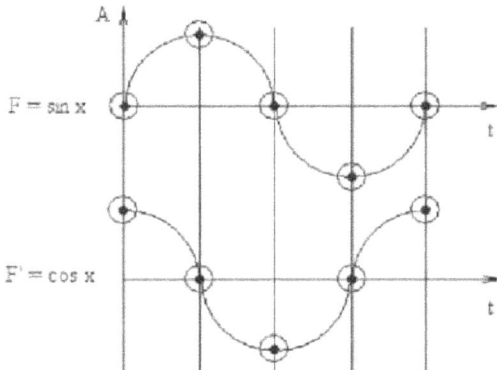

Fig. 33. Graphic differentiation of function sinX

From function/derivative relation it becomes apparent that first-order derivative extrema correspond to inflection points. In this case, when the function transits from negative to positive (or if we put it in another words, when the sign changes from minus to plus) the first-order derivative extremum will be positive. And vice versa, the extremum will be negative, in case the function changes the sign from positive to negative.

When using the graphic differentiation for research of compound signal, it is important to study the formation of first-order derivative extrema. Taking into account the fact that from physical point of view the derivative characterizes the rate of function variation, first-order derivative **curve extrema will show the limit of change of energy processes which form the function, namely first-order derivative maximum will characterize signal energy change from rise to deceleration, derivative minimum will characterize energy change from decrease to retardation. These boundary conditions are very important for dynamic processes. Blood flow is also a dynamic process; therefore any form of the biological signal of electrical or mechanical nature, which characterizes the cardiac rhythm, can be investigated using derivatives.**

107

ECG can also be investigated using graphic mathematical differentiation.

2.6.2. ECG as function of sequentially combined harmonic signals

If we consider ECG as a compound signal, we can notice that it consists of sequentially combined single-phase sinusoidal signals. Such composition is stipulated by maintaining the mechanism of blood flow pulsating oscillations which are shaped several times during one cardiac cycle by cardiovascular system. In this case, their amplitude with respect to each other is different.

Figure 34 shows ECG conditional model which consists of sequentially combined single-phase sinusoidal oscillations.

The main task of heart and vessels work is to provide normal blood flow. This work provides functioning of different parts of the heart and vessels in strict sequence, which is achieved due to harmonic oscillations of biological energy. The given oscillations have both amplitude build-up phase and attenuation phase. It is their boundaries that correspond to cardiac cycle phases. This theory is quite profound. It is assigned to the existence of blood superfluidity conditions which are maintained by the work of the entire cardiovascular system. Therefore, we shall only point out here that availability of ECG trace is quite sufficient to determine duration of cardiac cycle phases. Graphic differentiation can be quite successfully employed for determination of phases of energy oscillations which build up the biological potential of ECG.

Fig. 34. ECG conditional model which consists of sequentially combined single-phase sinusoidal oscillations

2.6.3. ECG graphic differentiation

Graphic differentiation principle is shown in Fig. 12. Compound signal, as ECG is, is differentiated in the same way. ECG curve does not allow to fix inflection points, but these points can be accurately plotted in local extrema on the derivative curve. **Figure 35 represents the curve of ECG trace and its first-order derivative. It can be seen that point P on the derivative, which is in the local extremum, corresponds to point P on the ECG. Point T is plotted in the same way. Point S is a significant one. In fact, there are no other methods which can be used for its plotting. The derivative allows to plot point S in the most accurate way. Thus, it enables to computerize the method of cardiac cycle phases measurement.**

The second derivative can be also used, but it makes no sense in this case, since the information value of the criteria is quite sufficient for tracing cardiac cycle phases using the first-order derivative.
When using CARDIOCODE instrument, the doctor constantly has to pay attention to the derivative. It helps him to check the accuracy of automatic processing of data. In case some error is traced, it will be possible to correct it manually.

Regular operations using the ECG database accumulated on the basis of multiple investigations help the doctor to be more confident in assessment of cardiac cycle phases.

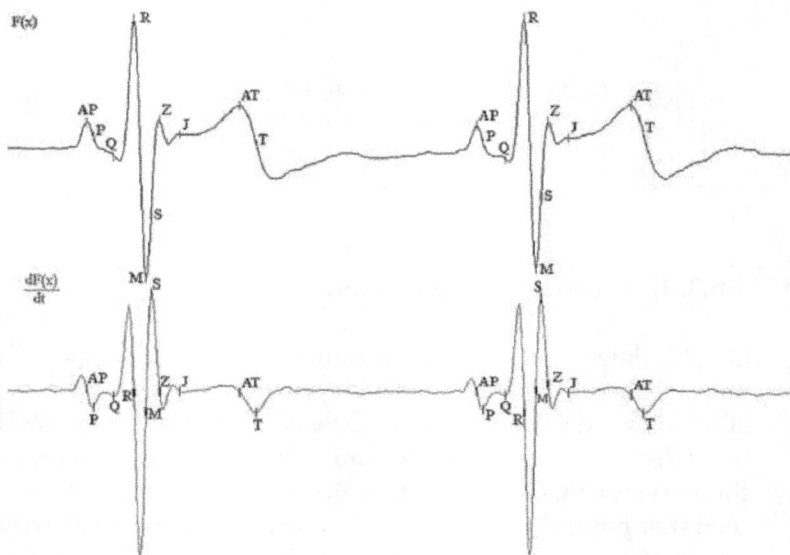

Fig. 35. ECG graphic differentiation

3. Criteria of hemodynamic parameters indirect measurement error evaluation using the elaborated method

3.1. General state of metrological support of indirect methods for hemodynamic parameters measurement

The goal of any metrological service is to ensure the uniformity of measurements. Unlike all other types of human activity, the medical science constitutes a major problem in reaching this goal. There are some basic aspects which should be taken into account to assure the uniformity of measurements. Each of these aspects should be thoroughly studied. These aspects are listed below:
a) standard instruments provision;
b) assurance of integrity of information obtained using the applied measurement techniques;
c) availability of methods for estimation of systematic, dynamic and static errors;
d) package of measures aimed at measuring instruments reliability assurance in terms of measurement errors stability.
There are also other constituents that affect the measurement uniformity system, but these constituents can be considered as derivatives of the aspects specified above. Therefore, let us consider the above-specified aspects in more detail.
In order to discuss the problems of the measurement metrology in the medical science in a reasonable way, it should be noted that the diagnostic parameters of the human body are divided into two categories: in vitro and in vivo. We have to accept this classification because in vitro category that includes such characteristics as height, weight, body capacity, etc. can be provided with standard instruments that are widely used in practice. The second category includes invasive parameters. It is rather difficult to talk about the

measurement of these parameters. Nowadays, these parameters can be only recorded and then estimated by the specialists. At present, there are no standard instruments that adequately reflect the basic features of the organs to be diagnosed and that generate biological signals in vivo mode.

The ways of solving the problem of signal recording in vivo can be found using indirect methods of measurement. This approach requires the development of theoretical models of the organ under study and the functions of this organ. The reliable estimation of the physiological parameters can be scarcely performed without having such models.

The simulation of the biophysical processes is greatly hindered by a number of problems. This allows some skeptics to speak about so-called "metrological dead-end"[135] [58]. To solve the simulation problem, it is necessary to have accurate data on organs functioning in the full range of norm and pathology. The main objective and intent of the full diagnostics of the human organism is to define the boundaries between the norm and pathology. However, it is impossible to simulate the biophysical processes without having measured the model parameters. It is necessary to have these parameters, even as hypothetical ones, but the required accuracy of measurement should be provided. In this case, any model shall undergo the process of check-out and adjustment of the parameters to be generated. This very focal process of simulation is the most difficult task. This problem has one further aspect. The necessity to allocate a lot of organism interrelations makes the process of determination of the basic features of human organism functioning very difficult. In this case, the abstracting process faces the problem of apriori recognition of a number of interrelated processes, since all human organism functions are interrelated and interdependent, whereas the organs have such a function as self-regulation. It is only clear for today that

135   Prischepa M.I. Features of the National Assurance of the Measurement Units in Cardilogical Diagnostics Laboratory [electronic resource]: Laboratory medicine. 2003 No.6. http://www.ramld.ru/articles/article.php?id=37 Screen title.

there is no needless thing in the human body, but this fact does not contribute to solving the metrology problem in the medical science. The necessity of the theoretical model development when using indirect measurement methods, stems from the situation occurred in the metrology field of the medical instrument making industry. However, there are various approaches to the simulation of the biophysical processes of the living organism. In general, these approaches can be defined as follows (the definitions provided below are developed by A. Prokopov[136] [59]:

1) identification of object features that are of great importance for the process of shaping the information signal.

2) identification of physical processes that cause the change of the information signal.

3) generation of equations with initial and boundary conditions.

4) quantitative analysis of the initial equations.

5) estimation of systematic error when using a measurement equation.

In this case, it is necessary to set apart the problem of metrological calibration of the instrument and estimation of the instrument influence on the distortion of the measured value.

The instrument can be calibrated using test-signals generated by the built-in generator. It is enough to perform instrument calibration. But the systematic error shall be estimated individually. Finally, the systematic error can be estimated using statistical method[137] [60] only.

The process of medical measuring systems development is described in the first section of the present paper. It should be added that this process can be considered as satisfactory if the comprehensive

136   Prokopov A.V. Algorithm of Justification of Equation of Measurement & Evaluation of a Methodical Error (Imprecision) in Indirect Measuring [electronicresource]:Access:http://mscsmQ.vniim.ru/files/2004/rus/prokopov-ru.pdf. Screen Title

137   Rudenko M.Yu. Development of Interferential Method in Measuring Arterial Blood Pressures & Instrumentation Based thereon. Thesis. Summary prepared by Mr.Rudenko M.Yu., Ph.D.. M., MNIIIMT, 1989, P.20.

investigation of this process in practice had shown minor deviations of the recorded values from the theoretical estimated values. These deviations will be considered as a systematic error.

The discussion of the absolute accuracy of these deviations is out of place in this case. It should be rather spoken about the reliability of these deviations. It is unreasonable to evince the truth by conducting the scheduled clinical trials only. The wide spread employment of the method and investigation of various specific states of the cardiovascular system by various specialists (using this method) will allow to refine the theoretical model that was taken as the basis for this method.

The development engineers and designers shall regularly publish the data on the estimation of the instrument static errors as well as data on achievements of diagnostics methodology development using the theoretical model, on the basis of which the instrument was designed.

The process specified above will lead to the development of a new paradigm in the medical sphere under study.

At present, the State Standard obliges the development engineers and designers to submit the procedures for primary and periodic calibrations of the medical measuring instruments for approval. Thereafter, the type of the measuring instrument is approved. Such a situation will persist for indefinite period.

Therefore, the special attention should be given to the issue of integrity of the recorded information. When designing the medical hardware, this issue should be studied in the wide range of the human organism states.

3.2. Sources of cardiac cycle phase transition parameters measurement errors occurring during ECG recording

It is reasonable to investigate the sources of errors that distort the integrity of information on the phase transition parameters in the recorded ECG taking into account two points.

The first point supposes the evaluation of compliance of the phase characteristics of existing standard leads with the elaborated method. This evaluation is to be performed using a common theoretical model for calculation of phase duration and hemodynamic parameters.

The second point supposes the evaluation of the electronic filters passband influence on the recorded ECG signal.

One of the objectives of this experiment is to determine the sources of the recorded information distortion. This allows to realize the causes of the crisis of phase analysis method development (this very crisis has hindered the solution of the measurement problem in cardiology earlier).

One patient was involved in this experiment, ECG was taken in several leads: in so-called HDA original lead developed within the frame of the present work (the term HDA is introduced by the authors, the abbreviation originated from hemodynamic analysis), and V3, V4, V5, V6 standard leads that are shown in figure 36.

Further, the duration of phases was calculated, and on the basis of these calculations, the hemodynamic characteristics were estimated. All the data were compared and verified to determine the static error of the deviations.

The most important objective was to determine the degree of integration and differentiation of the recorded ECG signals depending on the characteristics of the electronic filters.

The numerical data on the experiment results are provided in Table 2. ECG traces recorded in various leads are provided in Figures 37 to 40.

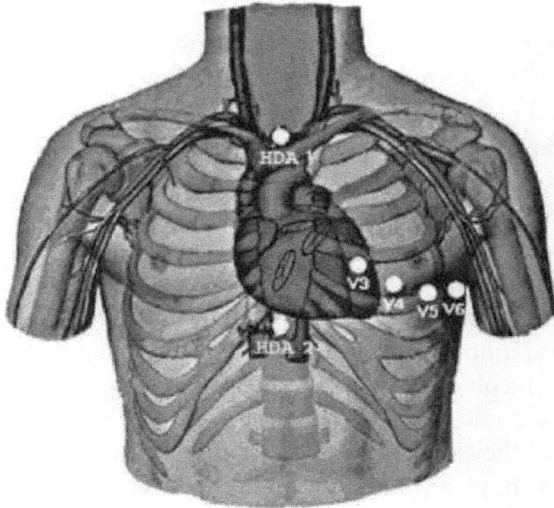

Fig. 36 Arrangement of ECG electrodes in V3, V4, V5, V6 standard leads and in newly developed HDA lead

Table 2.

Comparative data on the measured values in various leads. namely in **HDA** and **V3, V4, V5, V6** leads:

A) for equal filter passband values (provided that passband is the same as that used in HDA lead).

	QRS	RS	QT	PQ	TT	SV (ml)	MV (ml)	PV1 (ml)	PV2 (ml)	PV3 (ml)	PV4 (ml)	PV5 (ml)
HDA	0,093	0,047	0,363	0,084	1,013	72,4	4,3	50,7	21,7	42,9	29,4	10,7
V3	0,089	0,052	0,361	0,078	0,987	86,6	5,1	61,9	24,7	51,4	35,2	12,2
V4	0,088	0,043	0,355	0,088	0,984	62,9	3,8	43,4	19,6	35,2	25,6	9,6
V5	0,087	0,043	0,357	0,086	0,958	62,7	3,9	43,0	19,8	37,2	25,5	9,9
V6	0,090	0,046	0,341	0,096	0,961	69,3	4,3	46,6	22,6	41,1	28,2	9,9

B) for lower cutoff frequency equal to F_{lower} = 0.7 Hz.

	QRS	RS	QT	PQ	TT	SV (ml)	MV (ml)	PV1 (ml)	PV2 (ml)	PV3 (ml)	PV4 (ml)	PV5 (ml)
HDA	0,100	0,048	0,367	0,081	0,920	74,6	4,9	49,2	25,4	44,3	30,3	10,8
V3	0,114	0,053	0,383	0,061	0,814	87,8	6,5	54,2	33,6	52,1	35,7	12,2
V4	0,100	0,044	0,365	0,078	0,963	65,8	4,1	44,9	20,9	39,1	26,8	9,9
V5	0,089	0,046	0,343	0,093	0,888	68,0	4,6	43,9	24,1	40,3	27,7	9,9
V6	0,090	0,047	0,336	0,094	0,811	70,8	5,2	42,6	28,1	42,0	28,8	10,0

C) for lower cutoff frequency equal to F_{upper} = 11 Hz.

	QRS	RS	QT	PQ	TT	SV (ml)	MV (ml)	PV1 (ml)	PV2 (ml)	PV3 (ml)	PV4 (ml)	PV5 (ml)
HDA	0,123	0,070	0,378	0,070	0,970	133,9	8,3	93,2	40,7	79,5	54,4	15,8
V3	0,118	0,072	0,382	0,072	0,979	143,2	8,8	104,5	38,6	85,0	58,1	16,9
V4	0,128	0,072	0,379	0,063	0,998	141,0	8,5	100,8	40,2	83,8	57,2	16,2
V5	0,128	0,072	0,379	0,063	0,998	141,0	8,5	100,8	40,2	83,8	57,2	16,2
V6	0,127	0,072	0,358	0,075	0,863	136,3	9,5	88,0	48,4	81,0	55,3	14,9

D) for passband equal to F_{lower} = 0.7 Hz, F_{upper} = 11 Hz.

	QRS	RS	QT	PQ	TT	SV (ml)	MV (ml)	PV1 (ml)	PV2 (ml)	PV3 (ml)	PV4 (ml)	PV5 (ml)
HDA	0,138	0,069	0,401	0,050	0,964	134,7	8,4	92,3	42,4	80,0	54,7	16,2
V3	0,130	0,072	0,394	0,093	0,912	142,2	9,4	86,9	55,3	84,5	57,7	16,7
V4	0,130	0,072	0,373	0,068	0,843	140,2	10,0	88,9	51,2	83,3	56,9	15,8
V5	0,127	0,072	0,364	0,073	0,884	138,1	9,4	90,6	47,5	82,0	56,0	15,3
V6	0,130	0,074	0,357	0,080	0,833	140,1	10,1	86,1	54,0	83,3	56,8	15,0

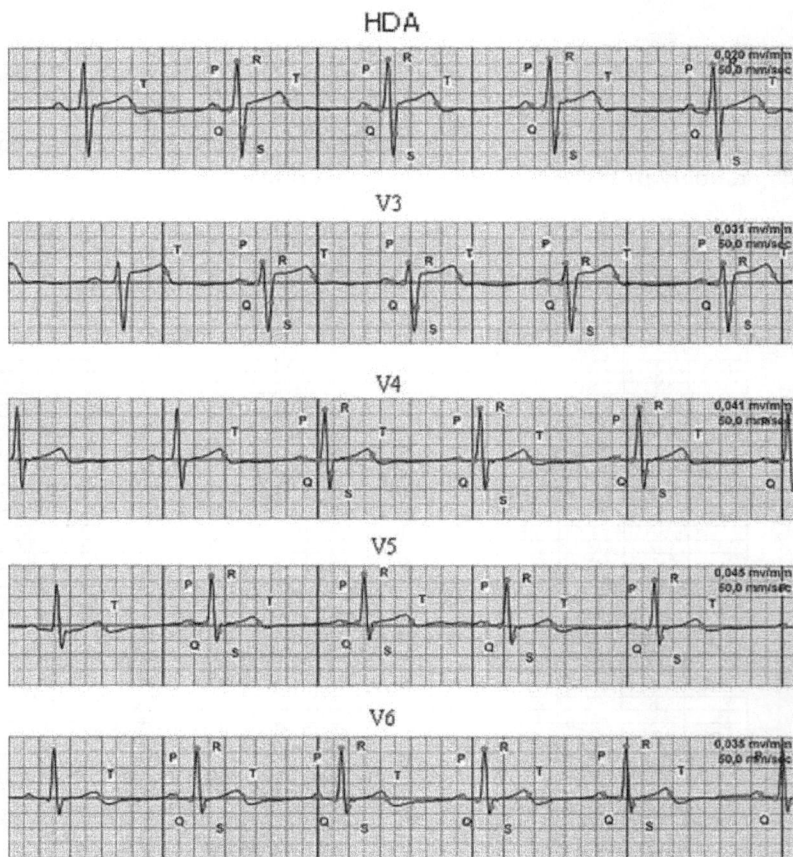

Fig. 37. Comparison of ECG traces with equal values of filter passband (that is the same as that used in HDA) in the following leads: HDA, V3, V4, V5, V6

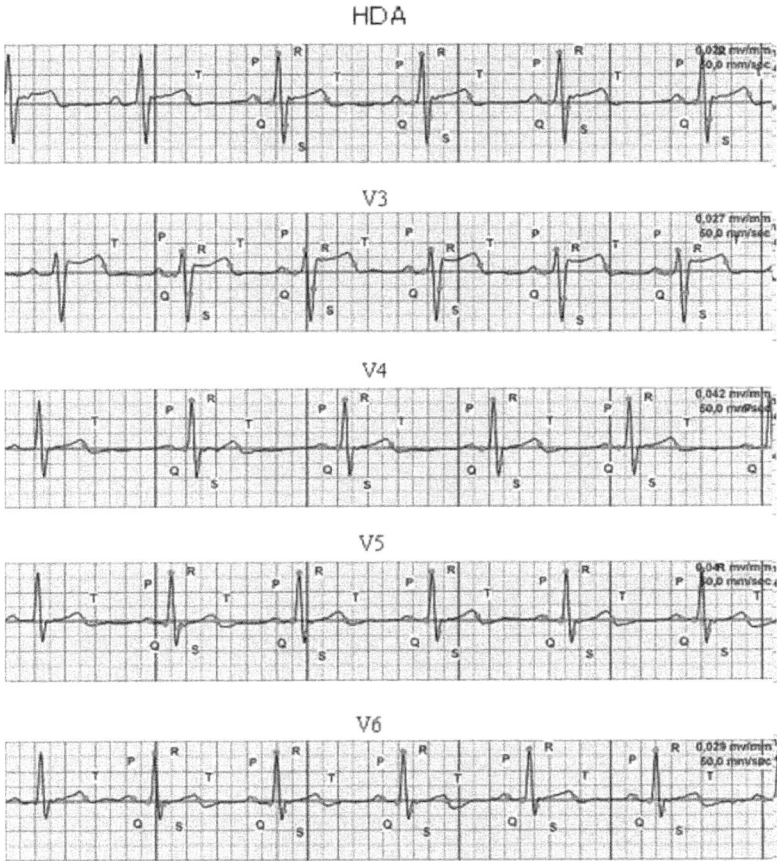

Fig. 38. Comparison of ECG traces (when lower cutoff frequency is equal to F_{lower} = 0.7 Hz) in the following leads: HDA, V3, V4, V5, V6

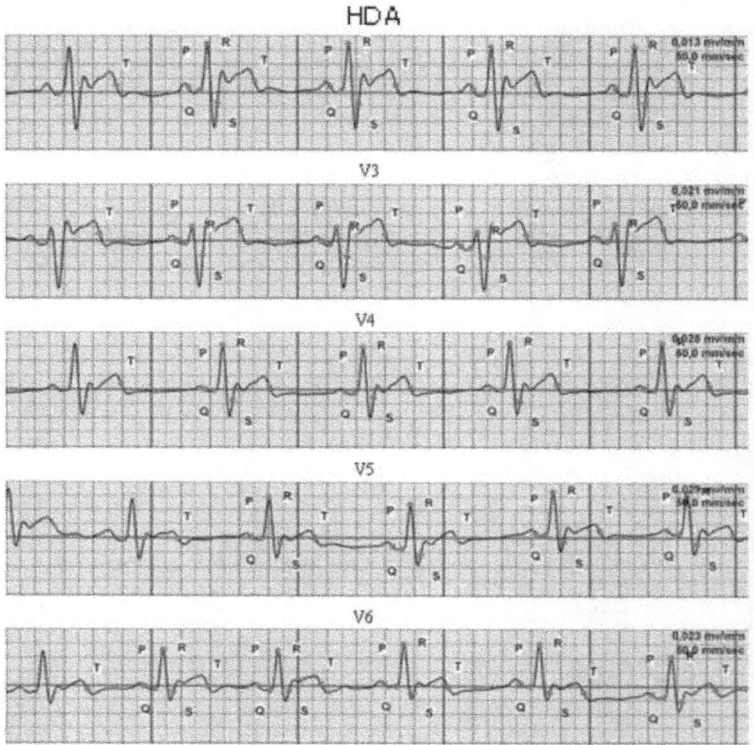

Fig. 39. Comparison of ECG traces (when upper cutoff frequency is equal to F_{upper} = 11Hz) in the following leads: HDA, V3, V4, V5, V6

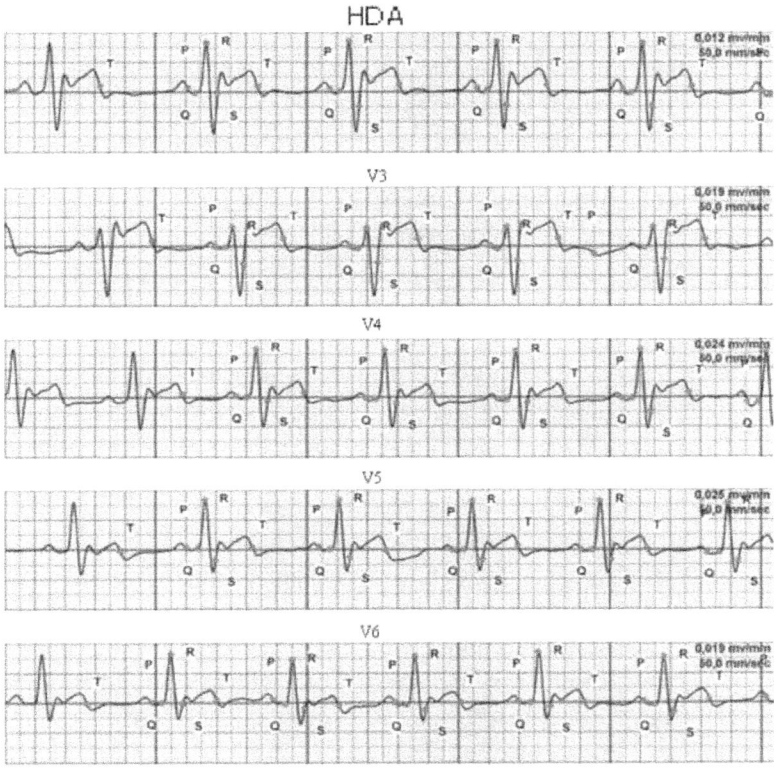

Fig. 40. Comparison of ECG traces (when passband is equal to $F_{lower} = 0.7$ Hz, $F_{upper} = 11$Hz)
in the following leads: HDA, V3, V4, V5, V6

The comparative analysis of the diagrams revealed the following basic features. HDA diagram is recorded when electrodes are arranged on the aorta and in the area of apex of heart. Using HAD diagram, the processes of cells depolarization over the whole area under study are recorded. Initially, V3, V4, V5, V6 diagrams were chosen to investigate the local areas. When comparing HDA and V3 diagrams provided in Fig. 37, it can be clearly seen that V3 lead does not trace the potential variations which control the aortic valve operation (contrary to HAD lead as is shown in Fig. 41). Moreover, the portion of diagram bearing the necessary information within the phase is smoothed.

As it is seen from V3 lead for RS phase shown in Fig. 41, the above-mentioned effect increases the duration of phase. This extra duration corresponds to three squares on the recorder calibration paper and two squares in HDA lead. It would seem to be an insignificant fact, but as it will be shown further, these very numerous ECG leads with incomplete amount of information on the phase transitions in each lead and with smoothing effect have resulted in the crisis of the whole ECG diagnostics theory. The tendency to use larger number of leads (to perform phase analysis) does not allow to record equal durations of the same phase in various synchronous leads. This tendency only increases the number of error sources. Figures 37 to 40 provide various versions of diagrams, on the basis of which the data presented in Table 2 were calculated. These diagrams demonstrate the problem of integration of information on phase transitions depending on the electrodes location.

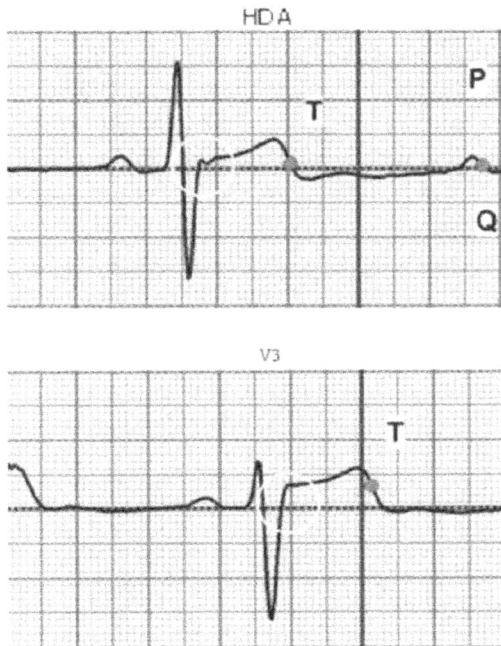

Fig. 41. Lack of information on the aortic valve operation in V3 standard lead and availability of this information in HDA lead result in the extension of RS phase in V3 due to the signal integration (smoothing)

Information presented above is true for all cardiac cycle phases.

G. Ivanov and co-authors have taken an attempt to rectify situati on in the theory of diagnostics using ECG by developing a high-resolution ECG method[138] [20]. Using the proposed method by performing the spectral analysis and further integration of the most informative harmonics, so-called late cardiac ventricle potentials were identified. In HDA method, these late cardiac ventricle po-

---

138   Ivanov G.G. High-Resolution Electrocardiography. M., Triada-X, 1999. P. 304.

123

tentials reflect operation of the aortic valve. The potentials are presented in

Fig. 41. However, the late cardiac ventricle potentials were recorded only to enable qualitative evaluation of these potentials by a doctor, but not to take any measurements of these potentials. In HDA method, the authors use the unified phase analysis theory (which describes only phase processes), but not abstract consideration of the phenomena like late cardiac ventricle potentials.

When speaking about the reliability of the values measured using HDA method, the feasibility of the fluid superfluidity conditions model developed by G. Poedintsev and O. Voronova[139] [14] should be confirmed.

Comparative data on quantitative measurements of stroke volume using HDA method and Teicholz method that is used in the standard ultrasonic investigation equipment.

		USI (ml)	HDA (ml)	Difference (ml)	Average value of difference (ml)
1	S.V.I.	171.3	173.5	2.2	
2	A.V.S.	59.4	58.5	-0.9	0.85
3	Zh.F.S.	58.6	66.7	8.1	(not more
4	B.P.F.	124.3	127.3	3.0	than 1% of
5	D.L.V.	198.6	200.9	2.3	compared val-
6	M.S.S.	101.8	99.7	-2.1	ues)
7	O.B.A.	79.3	78.0	-1.3	
8	Sh.V.T.	52.0	48.7	-3.3	
9	R.D.R.	95.2	94.9	-0.3	

This fact was also proved when performing comparative clinical trials of HDA method and widely-used Teicholz method (equation 1), that is applied to the ultrasonic investigation equipment (Table 3).

139   Voronova O.K. Development of Models & Algorithms of Automated Transport Function of the Cardiovascular System. Doctorate Thesis. Prepared by Mrs.O.K.Voronova, Ph. D., Voronezh, VGTU, 1995, 155 pages.

The difference between two methods is less than 1%. But HDA method allows to record values up to 0.001. It is more than enough for reliable diagnostics.

On the basis of data presented above, the results obtained using HDA method will be assumed as true values when performing further estimation of errors.

3.3. Estimation of hemodynamic parameter measurement errors using the elaborated method

Before passing to further processing of data provided in table 2, it is necessary to consider the proposed methodology, which is used to confirm the reliability of the elaborated method, and to estimate measurement result errors once again.

The essence of this methodology is in the artificial change of signal recording conditions (here we mean the signal which possesses informative attribute of the source information for hemodynamics equations), further estimation of percentage deviation from initial values, and detection of minimum deviations and the associated recording conditions on the principle of equality of blood volume flowing to and from the heart.

In this case we simultaneously analyze both elaborated HDA method and widelyused V3, V4, V5, V6 standard leads. Taking into account the fact that in each ECG signal under study there is simultaneous recording of several information criteria for determination of not one, but seven hemodynamic parameters at once, there appears a possibility to estimate percentage deviation from initial values of the whole spectrum of the measured information. In fact, we fail to estimate other errors, such as dynamic error, absolute error, arithmetic mean error and others, due to insufficiency of data.

At the last stage of the proposed method, we use the principle of balance of the blood volume flowing to the heart and flowing from the heart during one cardiac cycle. The object is as follows: determination of the whole spectrum of sources of signal distortion, and sorting out the signals and relevant recording conditions which are not acceptable for phase analysis. The total value of blood volume flowing to the heart and total value of blood volume flowing from the heart must be equal. Fig. 42 shows the philosophy of the proposed method.

126

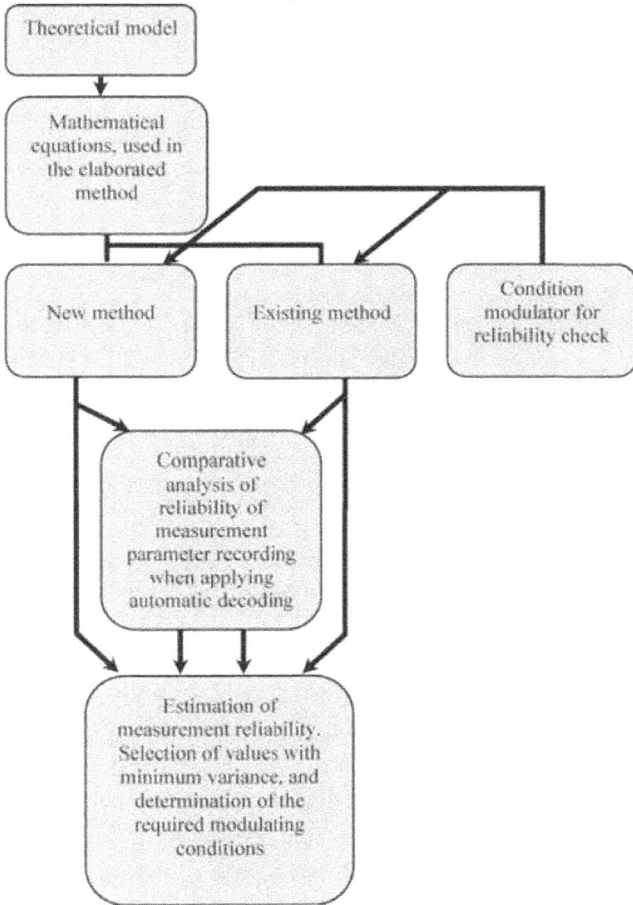

Fig. 42. Philosophy of recorded information reliability analysis and error sources determination

At the first stage of analysis, it is necessary to calculate the error percentage of static deviation from the assigned (as true ones) HDA values which are recorded with filter passband being the same as provided in item A of table 2. For the purpose of clarity, let us take them as a datum. In this case percentage deviation of the measured values presented in table 2 will correspond to the values presented in table 4.

Deviation percentage is calculated in compliance with the formula (20):

$$X = \frac{(N - N_{initial})}{N_{initial}} \cdot 100 \qquad (20)$$

wnere:

$N_{initial}$ — is the value which corresponds to A) conditions for HDA;

N — is measured value.

It is typical that when changing filter passband, in all ECG leads under study we observe the deviation from the true values both to positive and to negative. In case the passband is relatively narrow, the obtained data presented in Table 4 part D) differ by more than 100%. In HDA leads, the difference is more than 80%. The only conclusion we can come to is that the filter with parameters shown in Table 4 part D) deforms the signal heavily and can not be used in practice. We verified this conclusion at the second stage of our methodology and found out that the reason for this is a smoothing effect revealed when comparing the ECG traces in various leads (Fig. 41). The process of integration or differentiation is represented during signal passage through the filter, depending on signal amplitude and visual presentation of inflection points on signal curve. The quality of this process is dictated by harmonic frequency and amplitude, and it also depends on filter parameters. But it is not enough to determine conditional static error only. It can be seen from Table 4 that not all of the parameters vary equally relative to their initial value. So, in V6 lead (Table 4 part D), $\square$V2 parameter varies by 148% while PV5 parameter varies by 40.18%. For this reason, we should not use only one parameter to

128

evaluate the reliability of the estimated values. In addition to that, it is reasonable to use heart hemodynamic balance equation for data evaluation:

$$PV_1 + PV_2 = PV_3 + PV_4 \quad (21)$$

This is a physical constant which can not be violated in the mechanics of hemodynamic processes. It is reasonable to insert into formula the percentage variation provided in Table 4 instead of measured values provided in Table 2. The results are shown in Table 5.

Through detailed consideration of the equation, it is possible to determine the leads where the physical constant of balance (21) is violated least of all (regardless of true values deviation error).

Table 4

Percentage deviation of measured values in V3, V4, V5, V6 leads from values in HDA lead:

A) for equal filter passband values (provided that passband is the same as that used in HDA lead).

	QRS	RS	QT	PQ	TT	SV (%)	MV (%)	PV1 (%)	PV2 (%)	PV3 (%)	PV4 (%)	PV5 (%)
HDA	0.0	0.0	0.0	0.0	0.0	0.0	0.0	0.0	0.0	0.0	0.0	0.0
V3	-4.3	10.63	-0.55	-7.14	-3.06	19.61	23.25	22.09	13.82	19.81	19.72	14.00
V4	-5.37	-8.51	-2.20	4.76	-2.86	-13.12	-11.62	-14.39	-9.67	-13.05	-12.92	-10.28
V5	-6.35	-8.51	-3.03	3.38	-5.4	-13.39	-9.30	-15.18	-8.75	-13.28	-13.26	-11.21
V6	-3.22	-2.12	-6.06	14.28	-5.13	-4.28	0.0	-8.08	4.14	-4.19	-4.08	-7.47

B) for lower cutoff frequency equal to F_{lower} = 0.7 Hz.

	QRS	RS	QT	PQ	TT	SV (%)	MV (%)	PV1 (%)	PV2 (%)	PV3 (%)	PV4 (%)	PV5 (%)
HDA	7.52	2.10	1.10	-3.57	-9.18	3.03	13.95	-2.95	17.05	3.26	3.06	0.95
V3	22.58	12.76	5.50	-27.38	-19.64	21.27	51.56	6.90	54.83	21.44	21.42	14.01
V4	7.52	-8.51	0.55	-7.14	-4.93	-9.11	-4.65	-11.43	-3.68	-8.85	-8.84	-7.47
V5	-4.30	-2.12	-5.50	10.71	-12.33	-6.07	6.97	-13.41	11.05	-6.06	-5.78	-7.47
V6	-3.22	0.0	-7.43	11.90	-19.94	-2.20	20.93	-15.97	29.49	0.0	-2.04	-6.54

C) for lower cutoff frequency equal to F_{upper} = 11 Hz.

	QRS	RS	QT	PQ	TT	SV (%)	MV (%)	PV1 (%)	PV2 (%)	PV3 (%)	PV4 (%)	PV5 (%)
HDA	32.25	48.93	4.13	-16.66	-4.24	84.94	93.02	83.82	87.55	85.31	85.03	47.66
V3	26.88	53.19	5.23	-29.76	-3.35	97.79	104.65	106.11	78.00	98.35	97.61	57.94
V4	37.63	53.19	4.40	-25.00	-1.48	94.75	97.67	98.81	85.25	95.13	94.55	51.40
V5	37.63	53.19	4.40	-25.00	-1.48	94.75	97.67	98.81	85.25	95.13	94.55	51.40
V6	36.55	53.19	-1.37	-10.71	-14.80	88.25	120.93	73.57	123.04	88.81	88.09	39.25

D) for passband equal to F_{lower} = 0.7 Hz, F_{upper} = 11 Hz.

	QRS	RS	QT	PQ	TT	SV (%)	MV (%)	PV1 (%)	PV2 (%)	PV3 (%)	PV4 (%)	PV5 (%)
HDA	48.38	46.80	10.46	-29.76	-4.83	86.40	95.34	82.05	95.39	86.48	86.05	51.40
V3	39.78	53.19	8.53	10.71	-9.97	96.40	118.60	71.40	154.83	96.96	96.25	57.00
V4	39.78	53.19	2.75	-19.04	-16.78	93.64	32.55	75.34	135.94	94.17	93.53	47.66
V5	36.55	53.19	0.30	-13.09	12.73	90.74	18.60	78.69	118.89	91.14	90.47	42.99
V6	39.78	57.44	1.65	-4.76	-17.76	93.50	34.88	69.87	148.84	94.17	93.19	40.18

Table 5

Dynamic balance equation (PV1 + PV2 = PV3 + PV4) for calculated percentage provided in Table 4.

A) for equal filter passband values (provided that passband is the same as that used in HDA lead).

HDA	0.0
V3	3.62
V4	-1.91
V5	2.61
V6	-4.33

B) for lower cutoff frequency equal to F_{lower} = 0.7 Hz.

HDA	7.78
V3	-18.87
V4	-2.58
V5	-14.2
V6	11.48

C) for lower cutoff frequency equal to F_{upper} = 11 Hz.

HDA	-1.03
V3	11.85
V4	5.82
V5	5.82
V6	-19.71

D) for passband equal to F_{lower} = 0.7 Hz, F_{upper} = 11 Hz.

HDA	4.91
V3	-33.02
V4	-23.58
V5	-15.97
V6	-31.30

So,

for A) conditions, lead V4 has the least percentage deviation equal to minus 1.91%;
for B) conditions we determine the lead V4 deviation equal to minus 2.58%;
for C) and D) conditions we have HDA lead deviation equal to minus 1.03% and – 4.91%.
Based on Table 4, we can determine that the following leads have the minimum static deviation:

A) HDA and V6;
B) V4 and V6;
C) HDA, V4 and V6;
D) HDA.

However, in A) conditions V6 has the least static deviation for all parameters, one of which is 0.0%. But in this case, the principle of physiological balance is violated to a greater degree, i.e. minus 4.33% (Table 5).
We may also introduce a so-called reliability coefficient to the reliability estimation analysis:

$$K = K_1 \cdot K_2 \qquad (22)$$

where

K_1 – is the quantity of measured parameters;

$$K_2 = \frac{N_1}{1+N_2};$$

N_1 - is the number of normal measurements;
N_2 - is the number of faults.

It shows how efficiently the criteria of phase transition are revealed on ECG trace in automatic mode. It is very important for adequate consideration of individual features of a patient. But more profound investigation is required to identify the importance of each phase in each lead under various conditions within the scope of overall result of measurements.

Findings:

1). Filter passband is of paramount importance for reliability of ECG signal waveform recording.

2). Not all of the standard leads carry reliable information about all phases of cardiac cycle.

3). The elaborated method of ECG recording in HDA lead provides the most reliable information about cardiac cycle phase transitions.

4). V6 lead, which records ECG signal from apex of heart, is the closest to HDA lead of all standard leads with regard to the essence of information, but V6 lead signal reliability is lower.

5). Studies on reliability and diagnostic informativeness of HDA method with respect to standard methods must be continued in clinical conditions.

4. ECG recording using CARDIOCODE -1.
Criteria of phases registered on first-order derivative curve in automatic mode

4.1. Making the patient ready for ECG recording

The instrument shows high level of accuracy of cardiac cycle phases measurement. It registers the slightest changes in cardiovascular system condition including the patient emotional state which affects the hemodynamic processes. Bearing in mind this fact it is important not to provoke the patient agitation before ECG recording since this agitation can change the background of hemodynamic parameters. It is necessary to explain to the patient calmly the simple procedure of ECG recording to ensure the presentation of true hemodynamic parameters typical for real state of the patient. From our point of view, the doctor should possess the psychological skills and use some maneuvers to distract patient from diagnostic procedure which is to be made to attain a verdict with regard to his health. A word is a powerful means of influence on human emotional state, especially on emotional state of a sick person. That is why a word must be used very ably.

A patient is put on the bed, electrodes are set in appropriate points and then the instrument can be connected. The instrument is put on the bed near the patient, it is better to put it closer to the head. The emitting portion of the instrument is pointed at receiver placed at 1 to 2.5 m distance from the instrument. The instrument is connected with computer. The recording is effected after the instrument and computer are switched on.

4.2. Setting the electrodes for ECG recording.

Distinguishing features of single-channel ECG trace recorded using CARDIOCODE as compared with standard multi-channel ECG trace

Setting of ECG electrodes on the patient is a principle feature. Single-channel lead on ECG is a main factor which enables to minimize the measurement error. The electrodes are set as follows: one disposable electrode is set in the center of jugular notch near breast bone brim (Fig. 43). It is more efficient to localize this point using the needle therapy methods (acupuncture)[140] [61]. This location corresponds to VC 22 Tiantu. This point is interrelated with lungs work.

The second point is located on the middle front line of the body, namely in area where the branches of superior epigastric artery, branches of superior epigastric vein and medial cutaneous branch of anterior branches of seventh thoracic nerve are running. This is VC 14 Jugue point which is located on middle line 6 cune higher than umbilicus. 6 cune correspond to eight fingers which are put together tightly.

"Apparitor of the heart" is a literal translation of the name of this point. This point is used in needle therapy for strengthening the stamina of the heart[141] [61].

The third indifferent electrode is set near any other electrode (Fig. 43).

---

140   Havaa Luvson, Traditional and Current Aspectsof the Oriental Reflexotherapy. Edition 2, revised. M., "Nauka", 1990, 576 pages.
141   In the same reference. P. 139

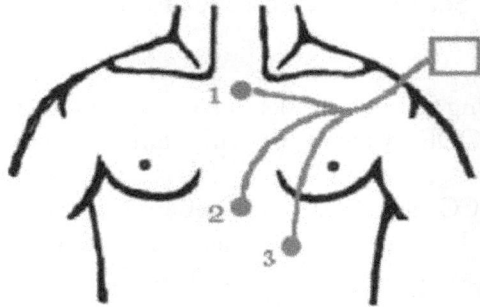

Fig. 43. Locations for electrodes setting.
Upper electrode is set in VC 22 point, lower electrode is set in VC14 point. The indifferent electrode is set in any convenient place

Both points are most informative for ECG recording when it is made for the purpose of cardiac cycle phases measurement. Let us see why the authors come to this conclusion.

To record the signals, the operational amplifiers are used. They are fitted with differential input. To be more precise, they record the difference of voltage on two inputs[142] [24]. If we fit the inputs on hands then the signal of standard first lead (upper extremity lead) is recorded. The signals of other leads, V3, V4, V5 etc., are obtained by the similar way.

Now we shall see, what information is registered when electrodes are fitted in points VC14 and VC22.

As is clear from Figure 36, the heart is located so that the first electrode is set in aorta area, the second electrode is set in myocardium area (to be more precise it is set in the area where the track of electrocardiosignal which enables the myocardium contraction ends). The amplifier will record the change of potential between these two points. Beginning from the instant when wave P is generated up to the instant when ORS complex is formed, namely in point S when myocardium is fully contracted,

142   Rutkovsky J. Integral Operational Amplifiers. M., Mir, 1978, P.180

the fullscale information (100%) is provided in the area between amplifier electrodes. The information is provided by amplifier without distortion. At that, one electrode, namely the electrode set in point VC22, is "grounded" virtually. That means that in this period the potential in point VC14 is changing relative to another point.

Beginning from point S, the aortic valve begins to open and the opening is in progress till point j is registered. These processes also occur in the area between the electrodes, full-scale (100%) information about them is also recorded by amplifier. At this, two inputs of amplifier are active and the resultant signal is a difference between two electrodes potentials. That is why the recording within the interval S–j is highly efficient. If even one electrode is shifted, then less than 100% of information is supplied into zone between the electrodes, the signal loss and distortion occur which is shown in figure 41 and figures 37 to 40.

Within the period from the instant of point j recording till the beginning of atrium contraction the signal is progressing in aorta. These processes also fully fall within the zone between amplifier inputs. The only difference is that the electrode in point VC14 is passive, i.e. it is virtually "grounded", and potential shift is recorded in point VC22.

In classical ECG the leads are obtained through the forced "grounding" of one electrode. This complicates the ECG trace recording procedure essentially. In addition to that, a great amount of information is lost in each lead.

That is why the authors have conducted serious investigations to obtain the answers to all questions which arose in the course of development of cardiac cycle phase analysis theory. To put it short, the authors investigated what was recorded earlier and what should be recorded.

4.3. Near-ideal ECG trace

Figure 44 shows the ECG trace and its derivative curve. This ECG trace is close to ideal. Periodicity of ECG cycles is clearly seen. All features of cardiac electrical activity are distinctly shown on the diagram. First of all, the potential variation at the instants of cardiac valves closing and opening is shown.

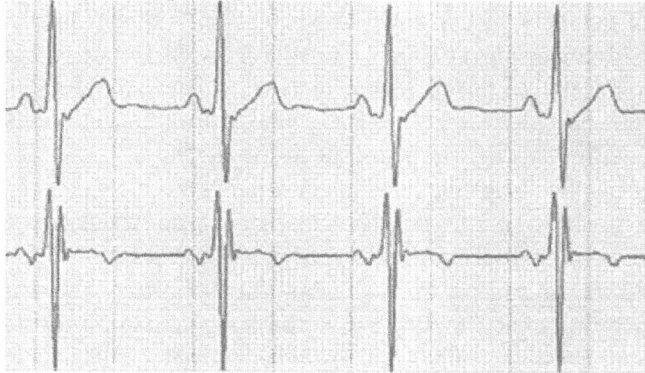

Fig. 44. ECG trace and its first-order derivative

The main phases of cardiac cycle are marked with dots as shown in Fig. 45.
The features of ECG trace and its derivative which shows the boundaries of cardiac cycle phases are described in more details further in the paper.

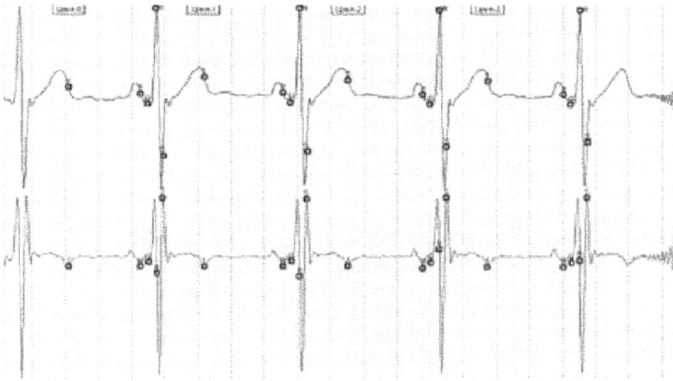

Fig. 45. Markup of cardiac cycle phases in first-order derivative extrima

4.4. Determination of phases based on ECG trace first-order derivative

4.4.1. Determination of point P

Point P on ECG trace corresponds to local minimum of derivative. In mathematics it is called an extremum of derivative. It is clearly shown in Fig. 46. On ECG trace the point of derivative extremum is represented by inflection point.

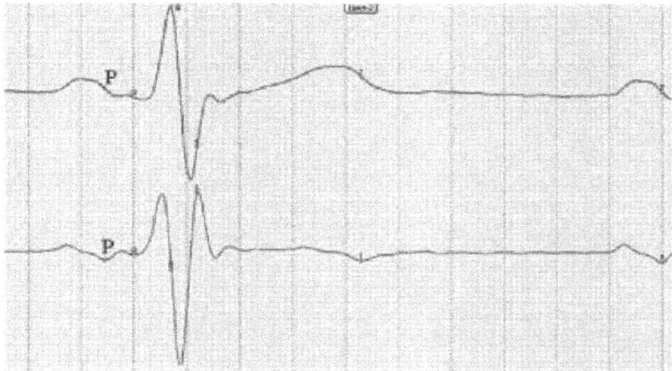

Fig. 46. On ECG trace, point P corresponds to inflection point of a descending curve of wave P, that corresponds to local minimum on ECG trace first-order derivative

It is impracticable to determine precisely the inflection point on ECG trace without use of derivative. This refers to any part of ECG trace. Atrioventricular valves start closing at the instant corresponding to point P. Amount of blood entering the ventricles decreases. At the instant corresponding to point Q, the valves are fully closed. The ventricles are filled with blood in amount required for build-up of the pressure necessary to eject the required quantity of blood into aorta.

Wave P represents the processes of rapid ejection of blood from atria. Each process comprises two stages. Rapid ejection begins at the instant corresponding to local maximum of derivative that corresponds to inflection point on leading edge of wave P on ECG trace. Further, the derivative curve has a declination to point P. Then the ventricles filling ends in valves closing at the moment corresponding to point Q.

Atrium is a major vessel. The atrium expansion performs the function of a pump aimed to build up the efficient pressure differential between periphery and the atrium ventricle. The efficiency of this work is high enough to enable the blood circulation throughout the body in any position of body and when the elasticity of vessels is lost. It is essential to note that point P can be easily determined both in normal and in pathological conditions.

4.4.2. Determination of point Q

Point Q, the same as point P, corresponds to another local minimum of the derivative. It is evident from Fig. 47 that point Q is located ahead of the QRS complex.

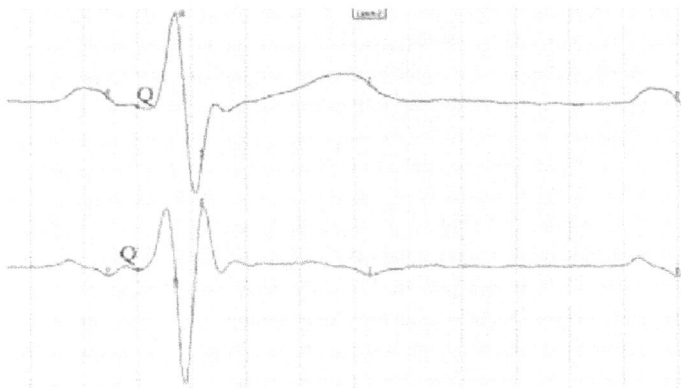

Fig. 47. Point Q determines the beginning of interventricular septum contraction. It corresponds to local minimum on ECG trace first-order derivative curve

Unlike point P, point Q is not always distinctly shown on derivative curve. If extremum is not clearly manifested by point Q, it can be distinguished by the instant of sharp escalation of derivative curve in this phase.

4.4.3. Determination of point S

Precise determination of point S is very important for hemodynamic parameters determination. This point is determined correctly in 99.9% of the whole amount of recorded ECG traces. This shows reliability and high confidence level of the method. Point S corresponds to maximum of derivative as shown in Fig. 48.

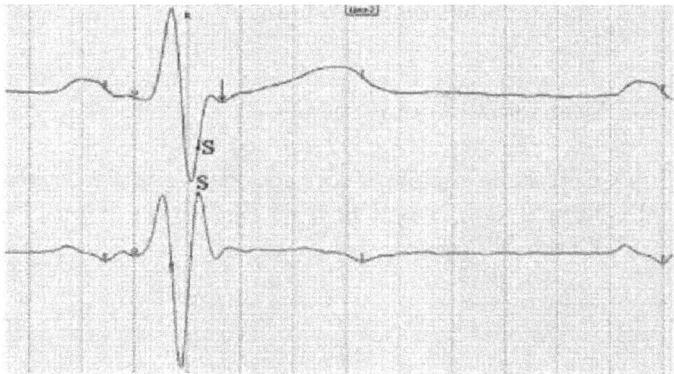

Fig. 48. Point S is determined by the most expressed maximum of derivative

At the instant corresponding to point S, the aorta valve begins to open. But the blood is still not supplied to aorta. In normal condition, the blood begins to enter the aorta after the point marked on the ECG trace with an arrow is passed.

4.4.4. Determination of interval S–j.

Analy sis of wave Z within this interval Interval S–j is very important for making the diagnosis. Within this period the aorta valve opens. This interval is not taken into account for hemodynamic characteristics calculation. It is used for qualitative assessment of work of aorta valve. The interval is recorded on ECG trace in the form of two waves. The wave with its top in point Z has a major amplitude. The amplitude of the other wave is much lesser. In normal condition, point Z must be located approximately at the level of isoline conventionally drawn from point P. At exercise load or in pathological state of the body, point Z can be considerably shifted upward or downward relative to isoline. As it has been already pointed out, the interval S–j consists of two waves. In norm, the blood begins to enter the aorta only beginning from the instant corresponding to point L which corresponds to onset of the second wave (Fig. 49).

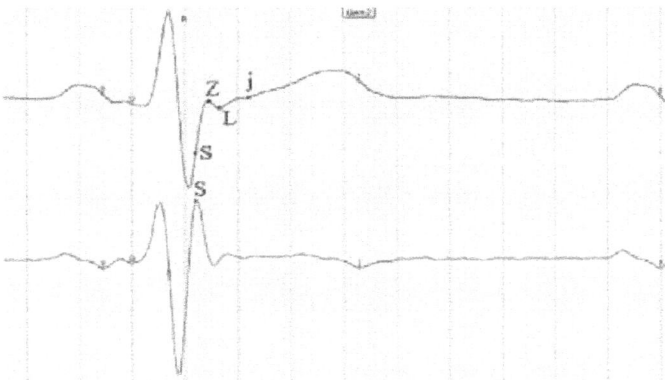

Fig. 49. In interval S–j, the aorta valve opens fully. Amplitude of wave Z specifies tension of valve opening

143

4.4.5. Determination of point T

Beginning from the instant corresponding to point j, the ECG trace shows the aorta work in the function of a pump. Wave T refers only to assessment of aorta work and is not related to myocardium work. Leading edge of wave T is very significant. It is originated in point j and has a plateau. It ends in point Tн. If the aorta elasticity is normal, the plateau is very short. If the aorta elasticity is lessened, the plateau is longer. At this, the other parameters can remain unchanged. This interval corresponds to slow ejection of blood from ventricles. The temporary expansion of interval depends on how quickly the volume of blood ejected from ventricle is distributed within aorta. The aorta, whose elasticity is normal, expands immediately and takes the shape corresponding to volume of blood. If elasticity is lower than normal, then some time is required for blood volume distribution along the aorta length. Wave T is generated only after blood is distributed within aorta. This instant is designated as T_H (Fig. 50).

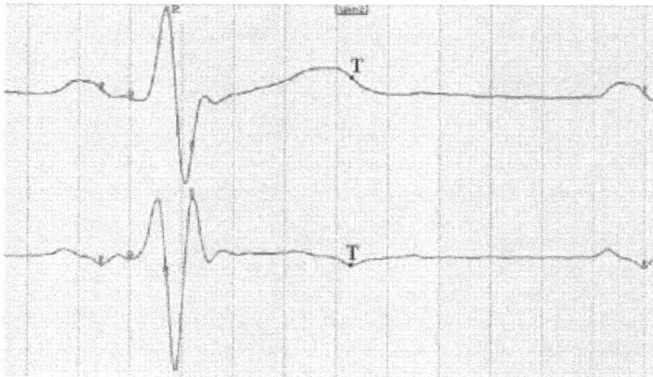

Fig. 50. Wave T is to be considered only for assessment of the aorta work and is not related to myocardium work. Length of interval j–T_H depends on aorta elasticity

4.4.6. Determination of point R

QRS complex determines the cardiac work in two-stage mode. First stage consists in interventricular septum contraction in phase QR. Second stage consists in ventricle wall contraction in phase RS. The interventricular septum remains contracted throughout the whole QRS interval (Fig. 51).

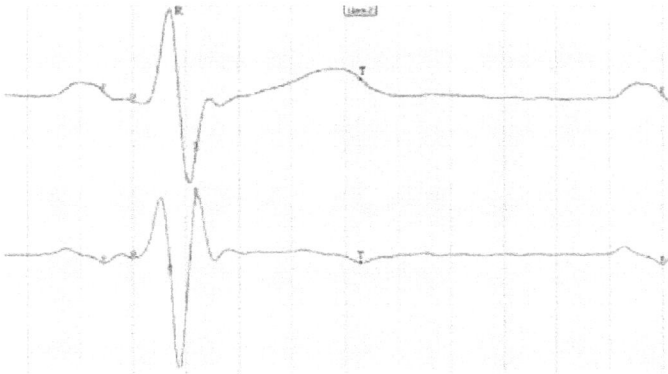

Fig.51. Point R is determined on ECG trace as a top of QRS complex similar as it is described in other scientific papers. This is the only point which does not correspond to criteria of phase registration based on derivative curve extreme

Myocardium tension is maintained till the instant of wave T end at which the aorta valve closes fully. Only after this the myocardium returns to its initial state.

5. Clinical practice of dynamics investigation on basis of ECG mathematical graphical differentiation

5.1. Volumetric parameters of hemodynamics measured by instruments

The method allows to diagnoze the following hemodynamics volumetric characteristics applying ECG single-channel lead:

SV – stroke volume of blood, ml;

MV – minute stroke volume, l;

PV1 – blood volume flowing to heart ventricle during diastole phase, ml;

PV2 – blood volume flowing to heart ventricle during atrial systole phase, ml;

PV3 – blood volume ejected by heart ventricle during rapid ejection phase, ml;

PV4 – blood volume ejected by heart ventricle during slow ejection phase, ml;

PV5 – blood volume (SV portion) transferred by ascending aorta acting like a peristaltic pump, ml.

146

5.2. Investigation of boundaries of physiological norm of cardiovascular system hemodynamics. Self-regulation of normal values of hemodynamic parameters

Investigations have shown that the heart possesses self-regulation and self-compensation mechanisms. Self-regulation mechanism is aimed at maintaining the normal hemodynamics of the healthy body on exertion. It determines the limits of the possible physical activity of a person.

Compensation mechanism provides minimization of myocardium local pathology effect on hemodynamic parameters. The mechanism implicates that the adjacent healthy area of myocardium starts working with double energy, thus compensating the cotractile hypofunction of pathologic area[143] [62].

Hemodynamic parameters are often similar during physical activity and in pathology. In both cases equal volume parameters may be recorded. In this connection it is important to have an answer to the following question: what effort on the part of cardiovascular system is required to maintain the norm?

Orthostatic test is very informative. When recording physiological parameters of cardiovascular system of a patient in the horizontal and immediately in upright positions, we may observe that natural blood pressure redistribution in the body is accompanied by substantial changes in the work of miocardium and vessels.

When cardiovascular system is healthy, the shapes of both ECG will not have substantial changes. In case of pathology, different ECG phases may have substantial changes. Comparison of these data enables to diagnoze actual capabilities of cardiovascular system. This fact is very important for the reliable prediction.

---

143   Phase Analysis of the Heart Cycle & Diagnostics Based on CARDIOCODE Application. M.Yu.Rudenko et al. Rostov-on-Don, Published by the Rostov State University, 2005, 56 pages.

Self-regulation and self-compensation mechanisms disguise the pathology to some degree, thus hindering diagnostics process. This may lead to the misdiagnosing.

Therefore, when considering pathologic case criteria on the ECG recorded by CARDIOCODE instrument, it is reasonable to start with the analysis of boundary states of the healthy organism, namely of a healthy patient who does not go in for sports, and an athlete exercising maximum physical load. Figure 52 shows ECG of two patients, one of them does not go in for sports, the other one is an athlete (weight-lifter).

It can be seen from Fig. 52a that ECG shape of the healthy patient has well-defined QRS complex and waves P and T. The associated hemodynamic parameters are shown in Fig. 52b. In the column specifying the percentage deviation from the norm (%) we can see that all parameters are within the norm, there are no deviations, i.e. they are equal to 0.0%. Individual boundaries of the norm are specified in the rightmost column. Thus, PV1 parameter, which is the volume of the blood flowing to the left ventricle during diastole phase, has the norm boundaries of 27.2 ml to 62.1 ml. The measured value (PV1 = 43.7 ml) is within the norm.

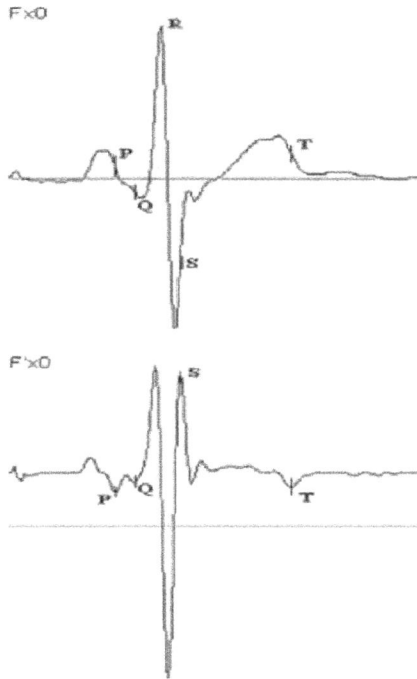

a) ECG of a healthy patient

SV (мл)	= 66,6 ½ = 0,0	43,3	100,0
MV (л)	= 5,0 ½ = 0,0	3,2	7,5
PV1 (мл)	= 43,7 ½ = 0,0	27,2	62,1
PV2 (мл)	= 22,9 ½ = 0,0	16,1	37,8
PV3 (мл)	= 39,5 ½ = 0,0	25,7	59,4
PV4 (мл)	= 27,1 ½ = 0,0	17,6	40,6
PV5 (мл)	= 9,8 ½ = 0,0	7,1	11,8

b) Hemodynamic parameters

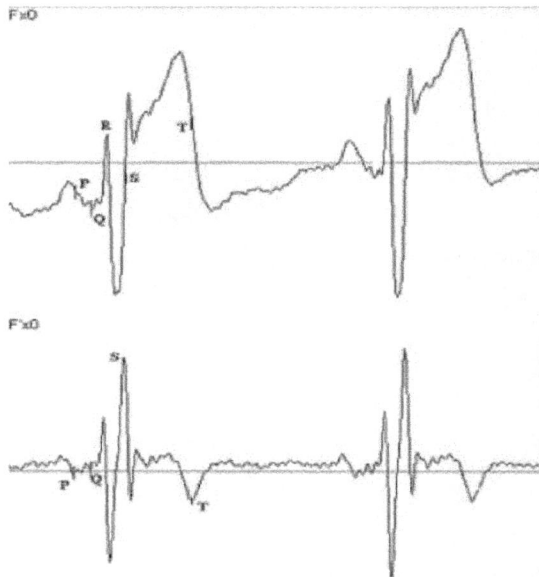

c) ECG of a weight-lifter (ECG was taken before training session)

SV (мл)	= 90,5	% = 0,0	45,5	107,7
MV (л)	= 5,2	% = 0,0	2,6	6,2
PV1 (мл)	= 71,0	% = 0,0	31,8	74,6
PV2 (мл)	= 19,5	% = 0,0	13,7	33,2
PV3 (мл)	= 53,8	% = 0,0	27,0	64,0
PV4 (мл)	= 36,7	% = 0,0	18,5	43,8
PV5 (мл)	= 10,5	% = 0,0	8,0	14,0

d) Hemodynamic parameters

Fig. 52 ECG of a healthy patient who does not go in for sports and ECG of a professional weight-lifter (with no exertion)

Figure 52b shows the ECG of a weightlifter (ECG was taken in no-exertion condition). It can be seen that the size of wave P is similar to the size of wave P on ECG trace of the patient who does not go in for sports. QRS complex has some distinguishing features. First of all, it is important to mention that the amplitude of complex lower portion (which characterizes contraction of ventricle walls in absolute magnitude) is larger than wave R. It is also evident that it is splited. Following point S, ECG portion corresponding to aortic valve opening period is characterized by the amplitude which exceeds wave R. Wave T amplitude is even higher. Wave T is responsible for aortic dilatation, it provides the work of absorption mechanism of aorta.

Hemodynamic parameters of the considered ECG are within absolute norm values (Fig. 52d). The only difference from those ones considered earlier is that individual boundaries of the norm, both lower and upper, are expanded. So, PV1 is equal to 31.8 ml and 74.6 ml respectively, if measured value is 71.0 ml. Paramenter percentage deviation from the norm is equal to 0.0%.

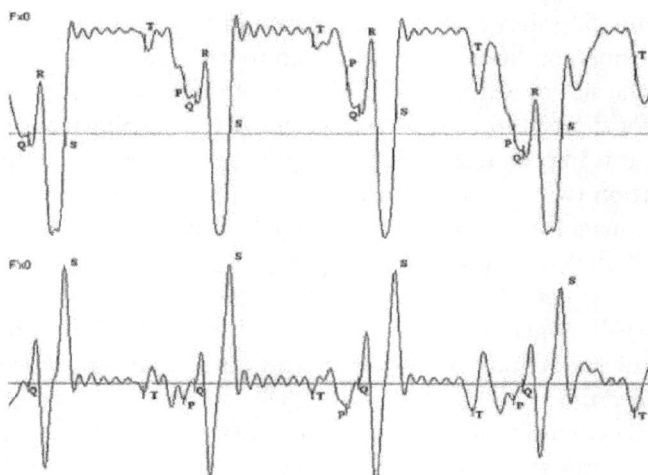

a) ECG of the weight-lifter (ECG was recorded within the period when the athlete lifted the barbell (weight 120 kg) and held it during 12 seconds)

SV (мл)	= 83,1 % = 13,4	35,7	73,2
MV (л)	= 13,4 % = 13,4	5,8	11,8
PV1 (мл)	= 275,8 % = -619,1	97,4	-53,1
PV2 (мл)	= -192,7% = 317,0	88,8	170,6
PV3 (мл)	= 49,4 % = 13,2	21,2	43,6
PV4 (мл)	= 33,7 % = 13,7	14,5	29,6
PV5 (мл)	= 8,7 % = 46,6	4,5	5,8

b) Hemodynamic parameters

Fig. 53. ECG of a healthy athlete in case of extreme exertion

152

Fig. 53a shows the ECG of the same athlete, but ECG was recorded half an hour after warming up exercises, when the athlete was lifting the barbell.

It can be seen that top of wave T is nothing but aortic valve oscillations. They modulate wave T. This can be observed only in the case when the vessels can not provide pumping function when transporting blood.

Table 6.

Relative physiological norms of cardiac cycle intervals (for rest state)

Interval	P-Q	R-S	QRS	Q-T	TT
ms	50...70	30...50	80...100	200...400	500...1000

The end of wave T and the beginning of wave P coincide. Such coincidence indicates insufficiency of cardiovascular system capability.

Waves T and P are equal in amplitude, they are very large. It happens in case of high tension of vessels when vessels perform the pumping function.

The splitted lower portion of QRS complex indicates the overexertion during ventricle walls contraction.

Hemodynamic parameters show the physical overexertion of the body to the full extent (Fig. 53b). It can be seen that all parameters have significant deviations from individual norm, particularly PV1 (blood volume flowing to left ventricle during diastole phase) and PV2 (blood volume flowing to left ventricle during atrium systole phase). So, PV1 is equal to minus 619.1% and PV2 is equal to minus 317.0%. The blood supply insufficiency is tremendous. It predetermines the ECG shape. In fact, the heart turns into the vessel, and pumping function is provided only due to the vibration of myocardium and aorta. Aorta is overstrained by 46.6%. We must point out, all parameters come to normal condition in a few minutes of calm after exertion.

We consider the patient to be ill if he needs (because of his physical health) to take hospital treatments due to 30% deviation from the norm of one or several hemodynamic parameters. On this background, hemodynamic parameters of the athlete on exertion demonstrate great capabilities of the organism.

Table 6 represents relative physiological norms for some of cardiac cycle intervals.

CARDIOCODE method allows to reveal the specific features of cardiovascular system which potentially may lead to hazard consequences in the form of severely disturbed blood circulation.

Hereafter, we shall consider the cardiac cycle phase duration and hemodynamic parameter values interrelation.

5.2.1. Phase duration variation effect on hemodynamic parameters

PQ phase decrement corresponds to decrement of time required for atrioventricular valve closing, it does not exercise essential influence on hemodynamic parameters. So, PV1 and PV2 remain within the norm. All the other parameters do not change.

PQ phase increment corresponds to increment of time required for atrioventricular valve closing, it does not exercise essential influence on hemodynamic parameters. Thus, if phase duration increases more than threefold, PV1 parameter decreases only to 1% of the normal value, PV2 parameter remains within the upper norm. All the other parameters do not change.

QT phase increment. Wave T, which is responsible for aorta pumping function, is a constituent of this phase. Wave T duration increment equal to 30% results in PV5 parameter deviation from the norm for some percents. This determines the aorta load.

QT phase decrement. If duration of wave T decreases, with QRS complex value being the same, then 30% decrement in duration is accompanied by RV5 rapid change (by 30%). This is the case of pronounced insufficiency. In this case, SV and MV decrease to the lower boundaries of the norm, PV2 also decreases by 20% of the norm.

RS phase increment of 20% with the unchanged Q and R causes PV1 parameters deviation up to + 5%. Although other parameters increase, they remain within the norm.

Wave R shift to the right or to the left in QRS complex causes proportional changes of all parameters. So, 25% shift of wave R to the right causes 25% deviation from the norm too. In a similar way the shift to the left causes changes of all parameters: they increase by up to 10%.

Next we shall consider dynamic changes of ECG shape in each cardiac cycle phase.

5.3. Research of cardiovascular system parameters with pathology in some cardiac cycle phases

5.3.1. Diagnostics of phase I and phase II (atrium filling/atrioventricular valve opening phase and atrium emptying/ventricular filling phase)

Change of wave P amplitude, the same as change of wave T amplitude, provides the pumping function of vessels. Wave P provides the additional atrium filling process. Atriums are major vessels, they have valves connecting them with low tension (venous) circulation system. Atrium enlargement provides pressure differential between atriums and the whole vascular system. In fact, this is a pump which successfully pumps out the blood grom vascular system irrespectively of its elasticity characteristics.

The case of the pathology with the pronounced deviations in hemodynamics is shown in Fig. 54a.

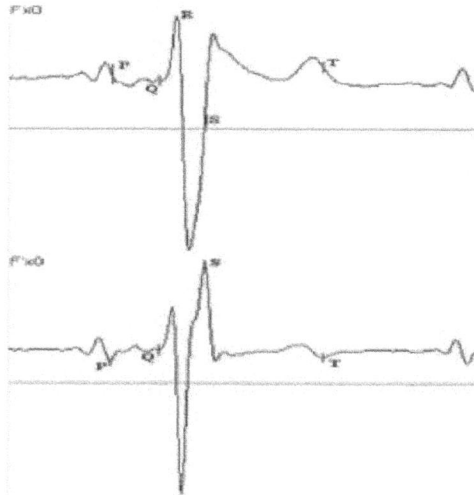

Fig. 54a. ECG trace of a patient with degenerative calcinosis of aortic valve with stenosis prevalence

156

SV (мл)	= 200,9 % = 86,7	45,5	107,8
MV (л)	= 11,6 % = 86,7	2,6	
PV1 (мл)	= 113,0 % = 52,0	31,7	74,4
PV2 (мл)	= 87,9 % = 164,4	13,8	33,2
PV3 (мл)	= 119,3 % = 86,7	27,0	68,9
PV4 (мл)	= 81,6 % = 86,7	18,5	45,7
PV5 (мл)	= 25,4 % = 81,0	7,9	14,0

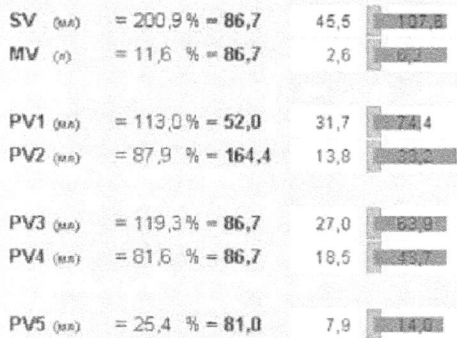

Fig. 54b. Hemodynamic parameters

5.3.2. Diagnostics of phase III (atrioven-tricular valve closing phase)

Atrioventricular valve closing begins after pressure in atrium drops. The signal which is controlled by this process is similar to the signal which controls aortic valve. On normal ECG trace, they are symmetrical with respect to QRS complex and almost identical. On ECG trace, represented in Fig. 55a, P–Q phase differs much from symmetrical S–j phase. The difference consists in the following: P–Q phase is U-shaped, while the S–j phase consists of two amplitude attenuation waves.

Regardless of specific shape of the phase, hemodynamic parameters do not deviate from the norm (Fig. 55b). It is known from practice that changes of this kind do not cause any symptoms of sickness and a patient feels well. Situation is a good deal worse if this U-shape is prolonged downwards, and the lower portion of QRS complex is drawn upwards. This shape manifests depleted resourses of myocardium and overloading of atrium. This is a very bad symptom.

As a rule, the diagnosis only states the fact of individual peculiarity of atrioventricu-lar valve closing, but it does not suggest a pathology case.

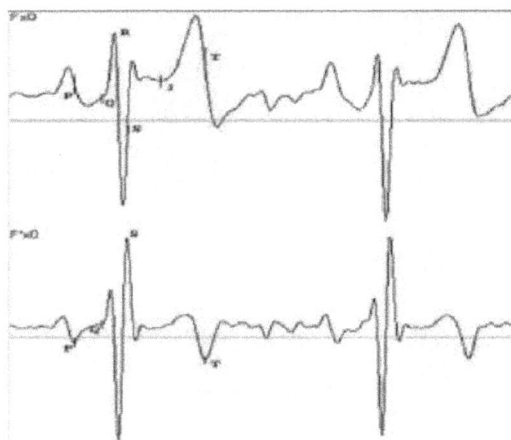

Fig. 55a. ECG of a healthy patient (65 years old)

SV (мл)	= 65,3	% – 0,0	44,4	103,9
MV (л)	= 4,3	% – 0,0	2,9	6,8
PV1 (мл)	= 42,9	% – 0,0	29,7	68,7
PV2 (мл)	= 22,4	% – 0,0	14,8	35,3
PV3 (мл)	= 38,7	% – 0,0	26,3	61,7
PV4 (мл)	= 26,6	% – 0,0	18,1	42,2
PV5 (мл)	= 9,9	% – 0,0	7,5	12,9

Fig. 55b. Hemodynamic parameters

158

5.3.3. Diagnostics of phase IV (interven-tricular septum contraction phase)

Interventricular septum contraction phase becomes more vulnerable to physiological changes as a body is aging. Fig. 56a shows that wave R amplitude is almost equal to zero. In some cases when using other methods of ECG recording, researchers erroneously come to a conclusion that wave R exists on S-side of the complex, and negative amplitude is nothing but Q-dip. We want to put emphasis on the inconsistency of this conclusion once again.

Hemodynamic parameters of the given patient are represented in Fig. 56b. It can be seen that all parameters, except for PV2, have significant deviations. PV1 has a maximum deviation, which is equal to 28.3%. This person has to take hospital treatment and to be under the care of physicians.

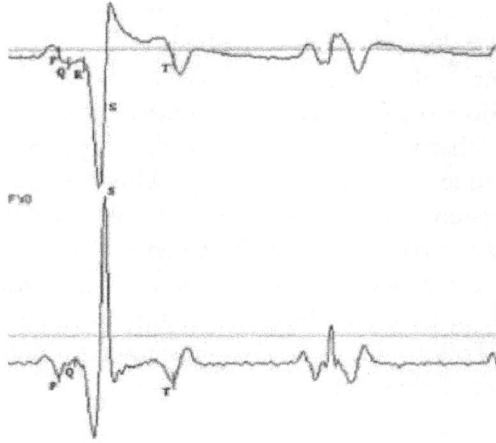

Fig. 56a. ECG trace of a male in-patient (78 years old) with postinfarction Cardiosclerosis .

SV (мл)	= 127,3 % = 18,7	45,4	107,2
MV (л)	= 7,5 % = 18,7	2,7	6,3
PV1 (мл)	= 94,6 % = 28,3	31,5	73,8
PV2 (мл)	= 32,7 % = 0,0	13,9	33,4
PV3 (мл)	= 75,6 % = 18,7	26,9	63,7
PV4 (мл)	= 51,7 % = 18,7	18,5	43,6
PV5 (мл)	= 16,6 % = 19,2	7,9	13,9

Fig. 56b Hemodynamic parameters

160

Figures 57 a and b show ECG trace and characteristics of a patient (64 years old) with postinfarction cardiosclerosis pathology.

The represented ECG manifests a total absence of interventricular septum contractility without errors. Hemodynamic parameters show necessity of immediate medical treatment of the patient.

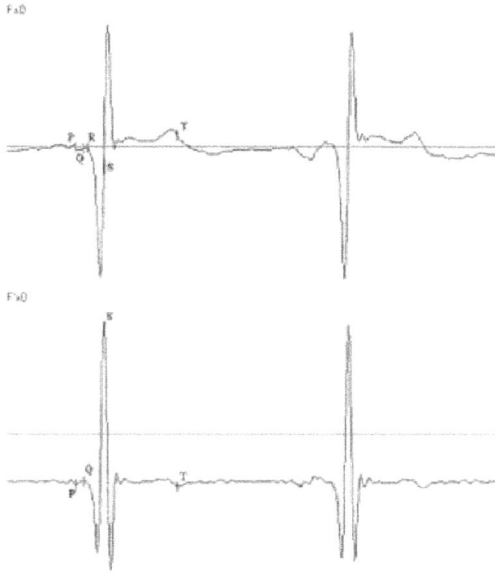

Fig. 57a. ECG without R wave

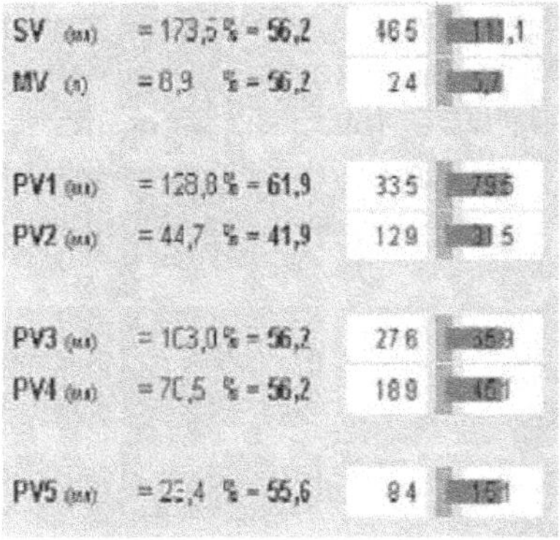

Fig.57b. Hemodynamic parameters

5.3.4. Diagnostics of phase V (ventricle walls contraction phase)

Figure 58a shows a typical shape of ECG trace which manifests actual absence of ventricle walls contraction. Patient diagnosis obtained using different methods of examination is as follows: postinfarction cardiosclerosis with chronic obstructive bronchitis. At this, hemodynamic parameters of the patient are within the norm (Fig.58b).

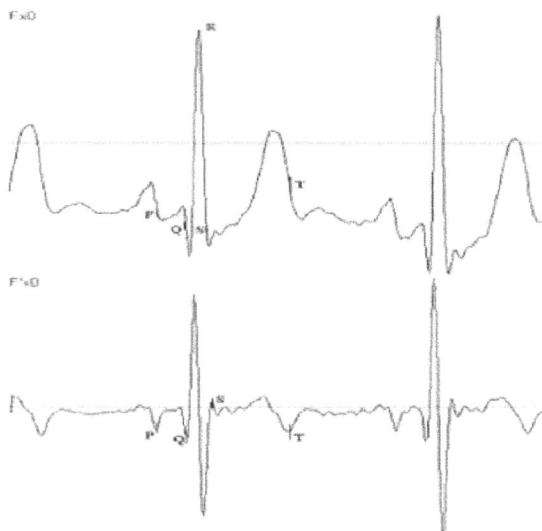

Fig. 58a. ECG trace of a male-patient (87 years old) with ventricle walls contraction function pathology

SV (мл)	= 67,2	% = 0,0	43,3	100 0
MV (л)	= 5,0	% = 0,0	3,2	7,5
PV1 (мл)	= 39,5	% = 0,0	27,3	62,2
PV2 (мл)	= 27,7	% = 0,0	16,1	37,8
PV3 (мл)	= 39,9	% = 0,0	25,7	59,4
PV4 (мл)	= 27,3	% = 0,0	17,6	40,6
PV5 (мл)	= 9,8	% = 0,0	7,1	11,9

Fig. 58b. Hemodynamic parameters

163

Diagnostic findings on the given cardiogram denote general myocardium weakness. Two factors denote this. Factor one: Q is extended downward, S is retracted upward. Factor two: wave T has a high amplitude, it proves the enhancement of aorta pumping function which compensates mi-ocardium weakness.

5.3.5. Diagnostics of phase VI and VII (semilunar valves tension and opening, rapid ventricular ejection phase)

This phase is the most significant one for the diagnostics. This phase is a link between miocardium and aorta. Its pathology does not introduce any major changes into hemodynamic parameters. But at the same time, ventricle – aorta isolation depends on its functioning. If the valve is weak and it allows the blood to flow to aorta in the process of ventricles contraction, this is the first reason of minimum arterial pressure growth. Finally, it results in hypertonia. ECG recording allows to diagnose this process indirectly. It is necessary to investigate amplitudes of two decaying waves following point S. If they are indistinct, it is necessary to examine the patient very carefully. Orthostatic test is required in this case.

If the amplitude of the first oscillation in S–j phase is high, it is possible to predetermine aortic valve elasticity loss which results in consumption of more energy for valve opening.

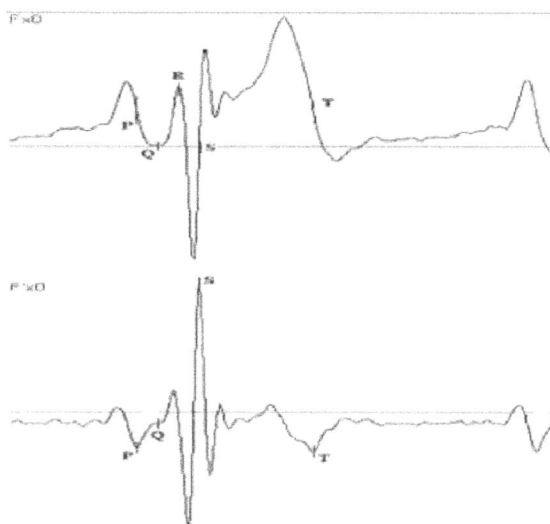

Fig. 59a. ECG trace of a male-patient man with aortic valve stenosis

SV (мл)	= 92,9	% = 0,0	45,8	108,6
MV (л)	= 5,2	% = 0,0	2,6	6,1
PV1 (мл)	= 68,2	% = 0,0	32,3	75,9
PV2 (мл)	= 24,7	% = 0,0	13,5	32,7
PV3 (мл)	= 56,1	% = 0,0	27,1	64,5
PV4 (мл)	= 37,8	% = 0,0	18,6	44,1
PV5 (мл)	= 13,2	% = 0,0	8,1	14,3

Fig. 59b. Hemodynamic parameters

165

5.3.6. Diagnostics of phase VIII (slow ventricular ejection phase)

This portion of the wave corresponds to the period between point j and point Tb, which corresponds to wave T beginning (Fig. 60a). Duration of this segment is much more important than shape. Blood volume is distributed along the length of aorta during this period. If aorta elasticity is sufficient, wave T will immediately be evident on the ECG trace. It indicates that the mechanism of aorta pumping function is normal. If aorta elasticity is low, time will be required for the distribution of the blood (supplied during rapid ejection phase) along the length of aorta. This time period is short, but this time is the main factor in vessels elasticity diagnostics.

This phase is related neither to the heart, nor to the vessels. It serves as a link between them. The mechanism of further behavior of waves T, U and P totally depends on the duration of this phase.

Hemodynamic parameters show that aorta is highly loaded, this loading exceeds the norm by 6.4% (Fig. 60b).

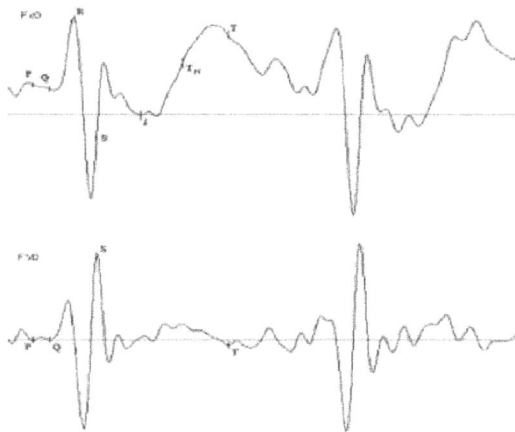

Fig. 60a. ECG of a male-patient (79 years old)

SV (ms)	= 78,0	% = 0,0	41,5	93,5
MV (л)	= 7,1	% = 0,0	3,8	8,5
PV1 (ms)	= 41,0	% = 0,0	22,5	50,0
PV2 (ms)	= 37,0	% = 0,0	18,9	43,5
PV3 (ms)	= 46,3	% = 0,0	24,6	55,8
PV4 (ms)	= 31,7	% = 0,0	16,9	37,9
PV5 (ms)	= 11,1	% = 8,7	6,4	10,2

Fig. 60b. Hemodynamic parameters

167

5.3.7. Diagnostics of phase IX (build-up of maximum systolic blood pressure in aorta)

This phase corresponds to the interval between the beginning of wave Tb and its end. It is determined by derivative ex-trema, which correspond to the inflection points of wave T leading edge and trailing edge (Fig. 61).

Wave T amplitude increment denotes the enhancement of aorta pumping function, which is caused by miocard load increment. As a rule, in this case hemodynamic parameters remain within the norm.

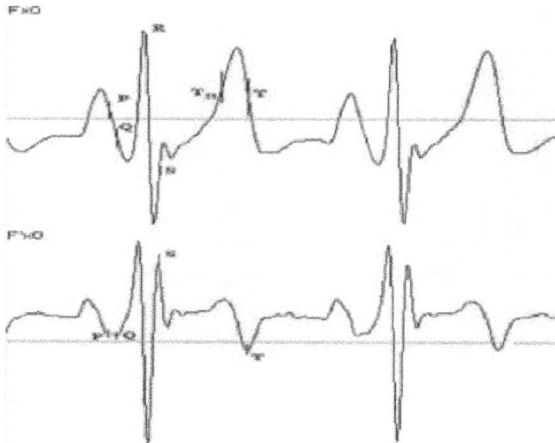

Fig. 61. ECG with increased T-wave amplitude

Smoothed leading edge of wave T indicates that aorta pumping function is weak. In this case physical exertion can result in unwanted sequela.

The existence of dip in wave T on ECG records taken from children arouses some interest (Fig. 62a).

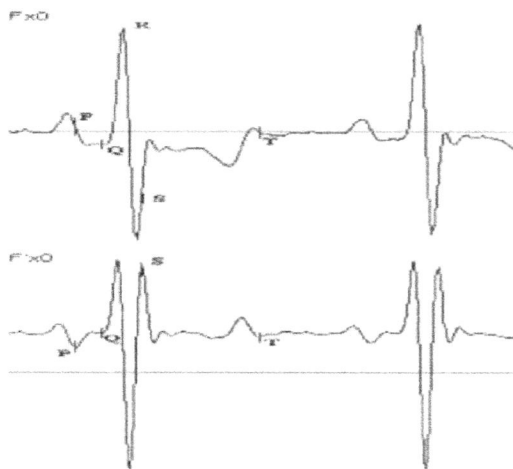

Fig. 62a. ECG trace of a child

SV (мл)	= 55,7	% = 0,0	40,8	91,3
MV (л)	= 5,4	% = 0,0	4,0	8,9
PV1 (мл)	= 29,6	% = 0,0	20,6	45,2
PV2 (мл)	= 26,1	% = 0,0	20,3	46,1
PV3 (мл)	= 33,1	% = 0,0	24,2	54,3
PV4 (мл)	= 22,6	% = 0,0	16,6	37,0
PV5 (мл)	= 7,5	% = 0,0	6,1	9,6

Fig. 62b. Hemodynamic parameters of a child

Investigations have shown that wave T has no dip. The peculiarities of the young organism provide normal elasticity of vessels and simultaneously do not allow to load the organism heavily with exertion in which case the wave T amplitude increases abruptly. Hemodynamic parameters are represented in Fig. 62b.

Similar dip of wave T can be observed on ECG traces taken from medical patients. This is due to the fact that aorta functional capabilities are depleted. Frequently it may be observed in postinfarction period (adults).

Figure 63 shows the shape of wave T which is similar to that shown in Fig. 87 (shape of wave T on ECG taken from the child).

As for middle aged patients, this dip is accompanied by pathologies. In figure 63a, wave R is also dipped. Wave P is distorted in ventricular filling phase. Pathology of atrioventricular valve is also observed. This is the case of extrasystole. Nevertheless, hemodynamic parameters are within the norm (Fig. 63b). The patient was taking hospital treatments at the moment when ECG was recorded. General state aggravated during motional activity.

5.3.8. Diagnostics of phase X (semilunar valves closing phase)

The given phase usually can not be diagnosed by ECG. It corresponds to the gap between point T, or we may also designate it a s Te (end of wave T), and point Ub (beginning of wave U). It is not significant from diagnostic point of view. Since the instant of valve closing, sanguimotion depends on venous system condition.

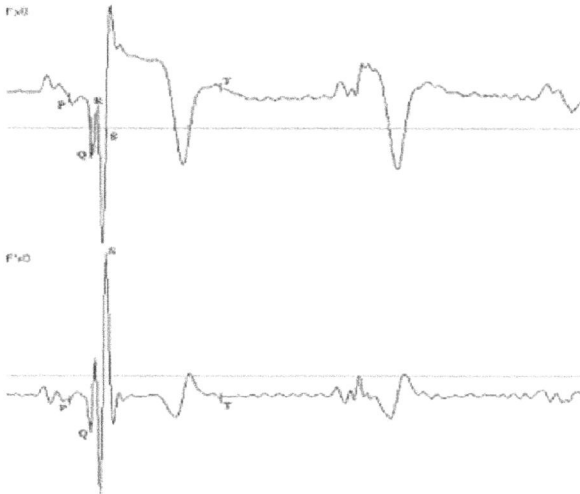

Fig. 63a. ECG trace of a male in-patient (81 years old)

SV (ил)	= 48,7	% = 0,0	45 8	108,8
MV (л)	= 2,7	% = 0,0	2 6	6,1
PV1 (ил)	= 35,3	% = 0,0	32 4	76,2
PV2 (ил)	= 13,4	% = 0,6	13 5	32,6
PV3 (ил)	= 28,8	% = 0,0	27 2	64,6
PV4 (ил)	= 19,8	% = 0,0	18 7	44,2
PV5 (ил)	= 9,9	% = 0,0	8 1	14,4

Fig. 63b. Hemodynamic parameters

171

5.3.9. Diagnostics of phase XI (premature diastole phase)

The tenth phase completes the cardiac cycle. This is a passive phase. During this phase, sanguimotion is provided by venous system. If venous system is in normal condition, then blood pressure gradient provides normal sanguimotion. If not, blood can inspissate, and blood pressure can exceed systolic pressure during venous circulation phase. The given processes can not be recorded distinctly on the ECG trace segment corresponding to this phase. Wave U is shown when myocardium can not perform its functions. If wave U is shown very distinctly, then pumping function of vascular system is in extremely critical condition which is represented by ECG trace in figure 64a.

Hemodynamic parameters differ insignificantly from normal values. But all physicians warn of risk to health. The woman herself has remarked, that physicians had told her that the men with such ECG would die in the nearest future, but the women could still live.

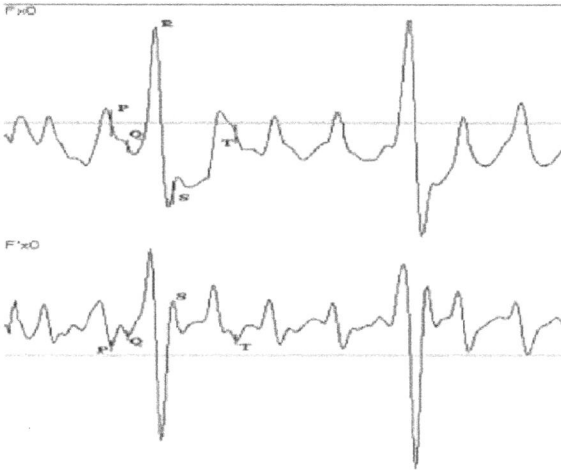

Fig. 64a. ECG of the woman who is not taking hospital treatment and keeps to regular life

SV (мл)	= 65,3	% = 0,0	42,4	96,8
MV (л)	= 5,4	% = 0,0	3,5	8,0
PV1 (мл)	= 24,4	% = -2,7	25,1	56,5
PV2 (мл)	= 40,9	% = 1,5	17,3	40,3
PV3 (мл)	= 38,8	% = 0,0	25,2	57,5
PV4 (мл)	= 26,5	% = 0,0	17,3	39,3
PV5 (мл)	= 8,5	% = 0,0	6,7	11,0

Fig. 64b. Hemodynamic parameters

173

5.4. Orthostatic test

We have considered the dynamic changes in each phase of cardiac cycle. We have highlighted the features which hinder the diagnostics, namely self-regulation and self-compensation mechanisms. But these features can be used to educe potentials of cardiovascular system, that is very important for more accurate prediction. For this purpose the orthostatic test can be made.

We can record ECG traces when a patient is in horizontal position and then without delay changes it to upright position. This records enable to reveal the pathologic phases that become evident due to blood pressure redistribution mechanism. This fact is very important, since the pathologic phases may not reveal themselves in one positions, and can be manifested to full extent in another position. The range of phase changing can be used as a criterion of criticality of cardiovascular system pathology. Figures 65 and 66 show the examples of orthostatic test.

It is characteristic that all phases may be subjected to change. Wave R amplitude changes most of all. Wave T is also subjected to change. Feeling of uneasiness was not present in horizontal position, in upright position the patient had a sensation of vertigo and dyspnea.

Hemodynamic parameters in both cases are within the norm. But when in the upright position, PV5 reaches the upper boundary of the norm only because normal parameters are reduced.

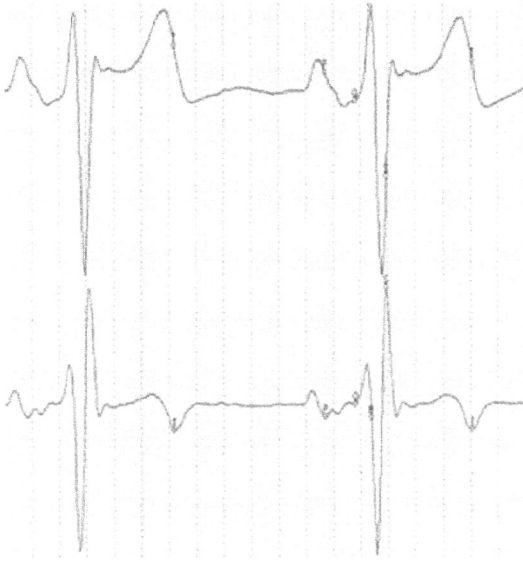

Fig. 65a. ECG trace, horizontal position

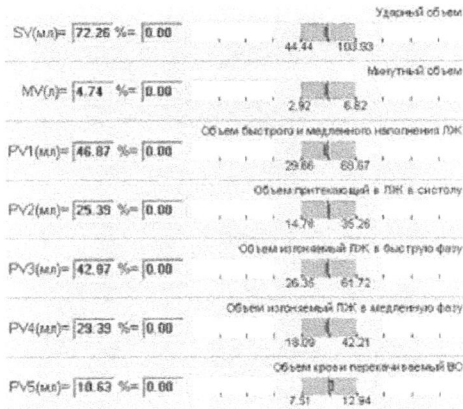

Fig. 65b. Hemodynamic paraments horizontal position

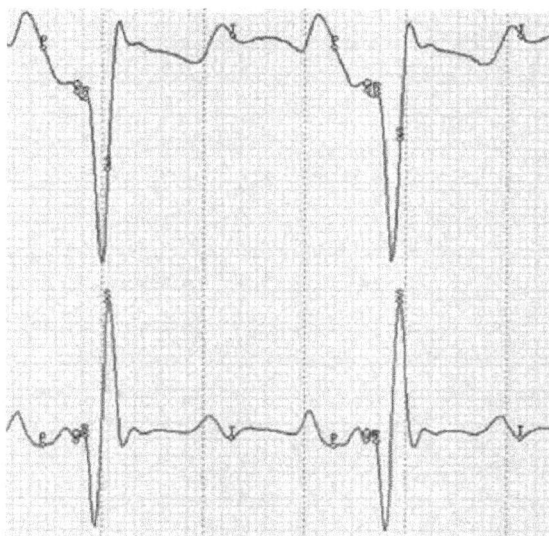

Fig. 66a. ECG trace, upright position

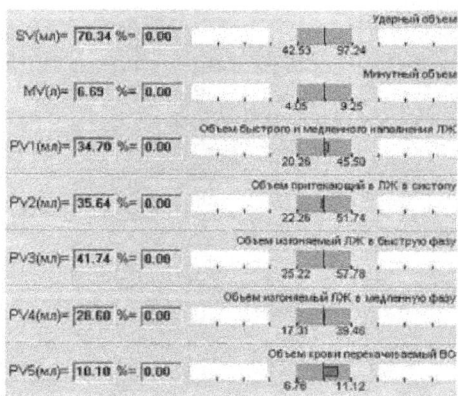

Fig. 66b. Hemodynamic paraments, upright position

176

6. ECG and rheogram recording using CARDIO-CODE-2

6.1. Electrodes arrangement procedure

It has been noted earlier that the rheogram is recorded using ECG electrodes. So we can specify the recorded rheogram as the dot rheogram. For this purpose, only two generating electrodes are incorporated into CARDIOCODE-2 instrument. These electrodes are arranged near ECG electrodes. They are connected to the similar standard disposable ECG electrodes. The electrodes arrangement is shown in Fig. 67. The data acquisition procedures are the same as those for CARDIOCODE -1 instrument.

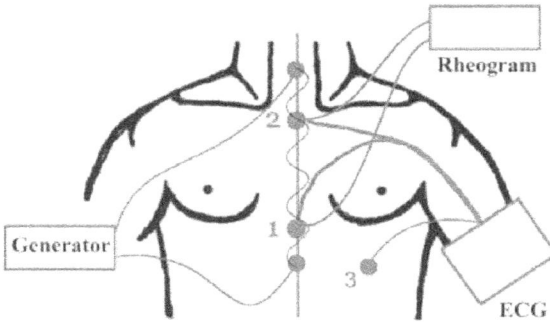

Fig. 67. Arrangement of electrodes for ECG and rheogram recording in the single point of body

6.1.1. Monitoring of ECG and rheogram recording

The distinctive feature of ECG and rheo-gram recording is minimization of respiratory undulation influence on the rheogram. If we can not minimize the respiratory undulation by bringing the patient into the comfort motionless position, then the patient should hold his breath while expiring slightly. The breath-holding time shall be 15 seconds maximum and shall correspond to the time period required for data recording. The breath should be held in the middle of exhalation to provide suitable conditions for recording of both signals within the whole recording period. All procedures are performed in signal viewing mode.

6.2. Rheogram phase characteristic

6.2.1. Phase mechanism of cardiovascular system functioning

To perform comparative analysis of the current potentials and vessel mechanical contractions in the given body segment, the difference between the processes carried in the human organism (in the central and peripheral blood circulatory systems) shall be taken into account. On the assumption of this condition, it is necessary to record ECG and rheogram in the single point using the common electrode. Recording of ECG and rheogram in different points of the body can not provide comparative information required for analysis of the processes carried in the given area.

The proposed method enabled to define the most informative areas where the ECG electrodes are to be arranged, and also enhanced reliability of the obtained data. As this method has been developed, the rheo-gram recording using ECG electrodes was enabled. Thereby, it became possible to determine the phase relations between ECG and rheogram in the common specific point of the body. Figure 68 provides the ECG and rheogram model that reflects the distinctive basic properties of the phase processes which take place in the ascending aorta.

As is clear from Fig. 68, there are seven distinctive phases in the given rheogram, while ECG provides a larger number of phases (i.e. ten phases). This is due to the fact that the phases provided on ECG and rheogram reflect different specific processes. These processes differ in their duration on ECG and rheogram. Therefore, the phases will be called variously.

The simultaneous analysis of the ECG and rheogram provides more information on the state of the cardiovascular system. This allows to improve the reliability of the diagnosis and to administer more effective therapy.

179

In general, the whole rheographic cycle can be divided into three segments. The first segment is the active work of the atriums myocardium, the second segment is time required for distribution of the stroke volume in the ascending aorta, and the third segment is the passive blood flow through the peripheral blood circulatory system. It is reasonable to begin the analysis from point S. In this point, the aortic pressure is equal to the diastolic pressure whose value is minimum in the given cycle. The importance of this point in the whole cycle lies in the fact that this point reflects the work of the aortic valve. In terms of mechanics, point S is the heart and blood vessels junction point. This very point carries the maximum mechanical load and is affected by the various pathological factors to a greater extend. The improper functioning of the aortic valve requires the continuous multiple treatment.

The aortic valve opening mechanism is complex. It consists of two periods, which can be observed on ECG trace as two decaying waves. On the rheogram, the first wave is presented by the level line, and the second wave has the specific rise of the curve. This very fact requires particular attention when performing ECG and rheo-gram analysis.

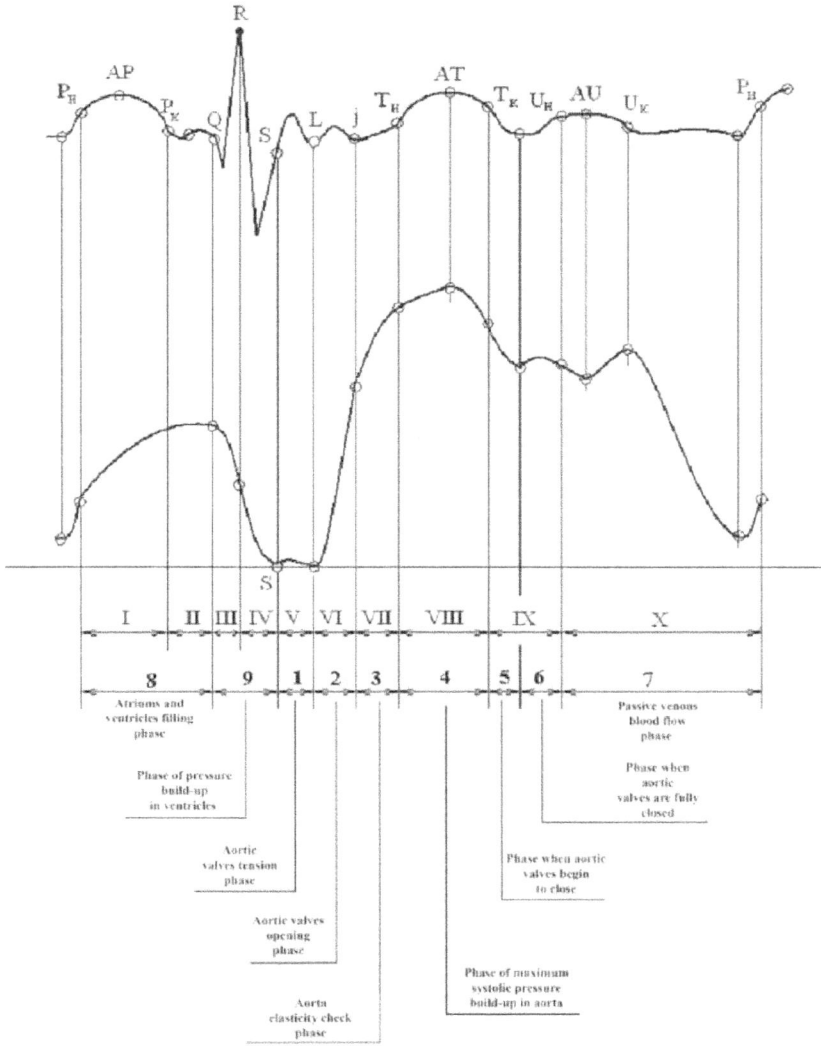

Fig. 68 ECG and rheogram phase correlation in ascending aorta (dot rheography: simultaneous recording of ECG and rheogram by taking readings from ECG electrodes)

181

The period of the systolic volume distribution through the ascending aorta begins immediately after the aortic valve is fully open. On ECG trace, this period is represented by the passive segment of the curve before the beginning of wave T. Only after the blood volume is distributed through the ascending aorta, the process of wave T generation begins. From the beginning of wave T to the beginning of wave P, the period of vessels contraction can be observed on ECG trace. On the rheogram, this period corresponds to the period of the blood pressure fluctuation in the major blood vessels and venous system.

The information presented above provides the pattern of the cardiovascular system phase work. Recording of the current potential in a single point and of mechanical contractions of aorta following this potential (particularly, in ascending aorta) allows to comprehend the blood superfluidity condition that is based on the blood flow pattern maintaining due to the pulsation of this blood flow. And finally, this allows to improve the accuracy and reliability of the diagnosis to a greater extend.

6.2.2. Aortic valve tension phase S–L

Let us consider ECG and rheogram that were really recorded using CARDIOCO-DE-2. For this purpose, ECG and rheo-gram of a young healthy man were recorded (Fig. 69).

It is reasonable to begin the consideration of the process of aortic pressure fluctuation from point S. On ECG trace, point S corresponds to the derivative extremum. The arising wave S–L on ECG trace corresponds to the almost horizontal segment on the rheogram. From physical point of view, it is due to the fact that the aortic valve is still closed, and blood can not be supplied to aorta. Therefore, the pressure does not rise on the rheogram. In case of valve pathology, the valve allows the blood to pass into aorta when QRS complex begins to develop. In this case, the pressure build-up will

182

be recorded in segment S–L. In greater detail, this process will be considered in the Section devoted to the diagnosing procedures.

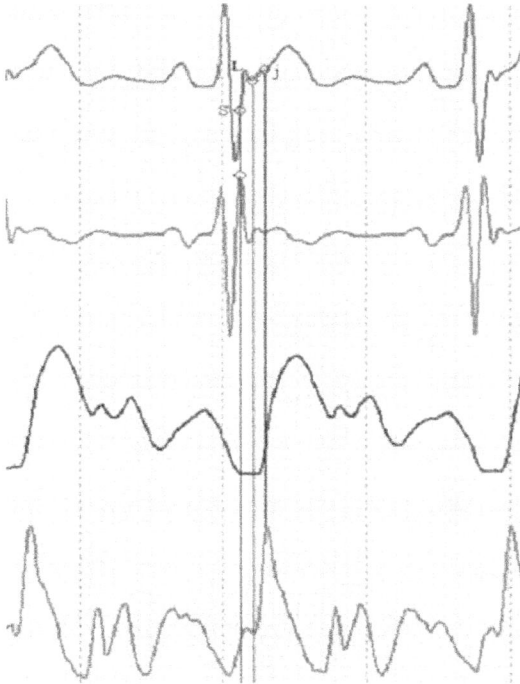

Fig. 69. ECG and rheogram of a young healthy man (ECG, its derivative, rheogram and its derivative curve)

6.2.3. Aortic valve opening phase L–j

The valve opens in the segment following the point L, and aortic pressure begins to rise sharply due to systolic discharge of blood to the aorta. The pressure rises sharply up to point j. On ECG trace, this period is represented by wave L–j. On the rheogram, point j can be seen very clearly. This point corresponds to the rheogram derivative maximum (see Fig. 69). In norm, the derivative extremum has one maximum. In case of valve pathology, the derivative extremum has two or even three maxima.

As viewed in the large, the whole period of aortic valve opening corresponds to phase S–j. Taking into account the importance of this period for cardiovascular system activity, this period should be investigated on rheogram part by part as it was noted above. This allows obtaining of very important detailed information on the aortic valve state.

6.2.4. Slow ejection interval (check of aorta elasticity) j to T$_{\text{н}}$

The given phase is related neither to the heart nor to the vessels functioning. This phase is the physiological pause, during which the systolic volume supplied to the aorta shall be distributed through the whole aorta. The extending pressure rise is observed on the rheogram. This process is represented in Fig. 70.

184

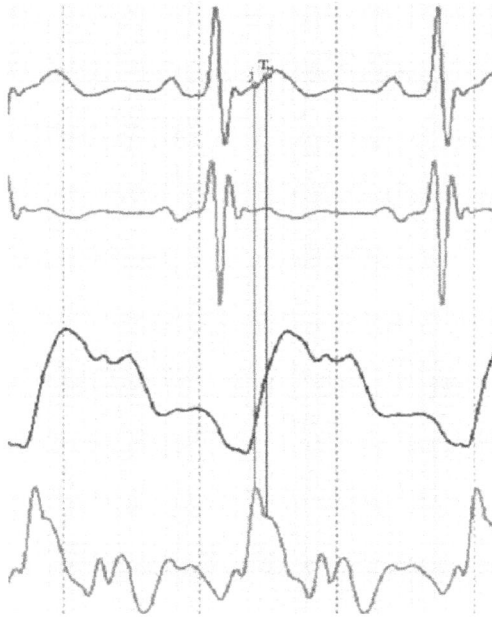

Fig. 70. Phase of aorta elasticity (compliance) check j–T$_{н}$

In normal condition, the pressure rise in this phase shall be 1/3 of the total pressure rise. The derivative shall have straight decaying segment. Following this segment, the pressure rise shall be observed. The process of the systolic volume distribution through the ascending aorta, depending on aorta elasticity, is shown in Fig. 71.

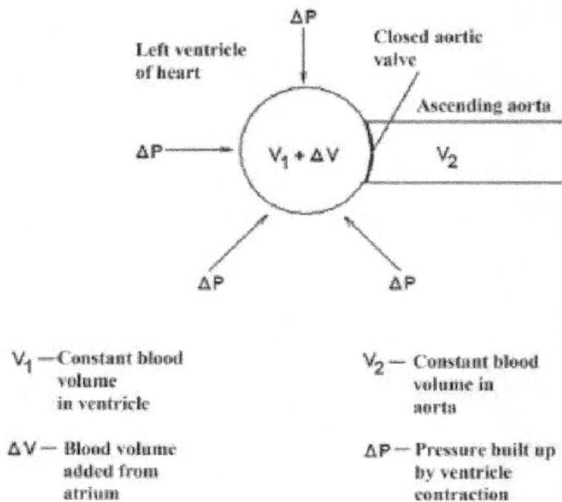

V₁ —Constant blood
volume
in ventricle

V₂ — Constant blood
volume in
aorta

ΔV — Blood volume
added from
atrium

ΔP — Pressure built up
by ventricle
contraction

a)

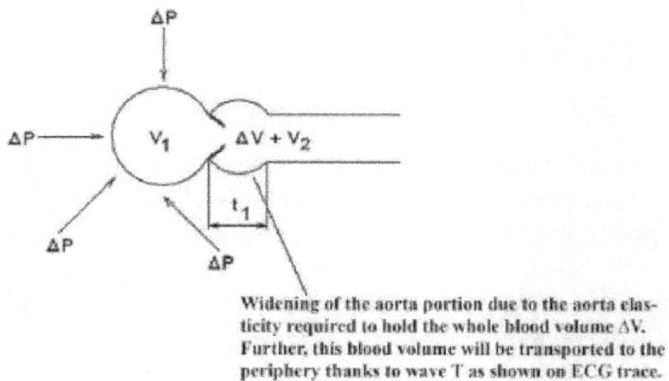

Widening of the aorta portion due to the aorta elasticity required to hold the whole blood volume ΔV.
Further, this blood volume will be transported to the
periphery thanks to wave T as shown on ECG trace.

b)

Fig. 71. Systolic volume distribution in case of normal aorta elasticity. The aorta
shall widen when affected by blood volume. The aorta distension depends on its
own elasticity. In case of proper aorta elasticity, the aorta shall distend (widen)
and assume the shape of the blood volume supplied from the ventricle

186

In case of normal aorta elasticity, the blood volume supplied to the aorta shall expand this aorta. This process depends on the aorta elasticity only. In case of the insufficient aorta compliance (this may occur due to the vessels sclerosis or other pathological factors), the blood volume has to be distributed along the whole aorta length. And this process requires more time than in case when aorta widening is normal (Fig. 72). On ECG trace, this process will be represented by the prolongation of phase j–Tн. On the rheogram, this process is also shown as the prolongation of phase j–Tн, but at the end of this phase, the pressure will gain its maximum (systolic) value. Thus, the pressure will not rise during the next phase. It will be represented by bifurcated vertex of the rheogram. This process, which is taken as diagnostic criterion, is described in the Section devoted to the diagnosing procedures. Let us only note again that phase j–Tн is related neither to the valve control mechanism nor to the vessels. On ECG trace of a child, this phase has the negative inclination angle. Some doctors talk about "wave T dip". But if we consider the grown-up, dip on wave T may indicate the pathology. This issue is covered in greater detail in Section devoted to the diagnosing procedures.

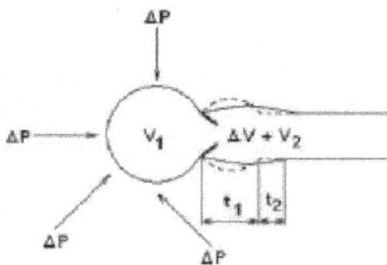

Widening of aorta along the amplitude due to reduced aorta elasticity will be less, while the time required for distribution of blood volume ΔV through the aorta length will increase by t₂.

Fig. 72. Systolic volume distribution when aorta elasticity is reduced. Aorta failed to distend (widen) and to assume the shape required for the volume of blood supplied from ventricles. But since the blood volume still must be distributed within aorta, extra time t_2 is required

6.2.5. Phase $T_н$ —$T_к$ when maximum systolic pressure is built up in aorta

In the next phase, the contractile function of vessels is activated. In previous phase, the systolic volume of blood has been distributed through aorta, and now the blood suction mechanism is activated to provide blood running through the vessels. We have already described the condition of blood superfluidity in vessels. This condition is provided by generation of wave T which is shown on ECG trace. In normal condition, maximum of wave T coincides with pressure maximum on the rheogram (see Fig. 73).

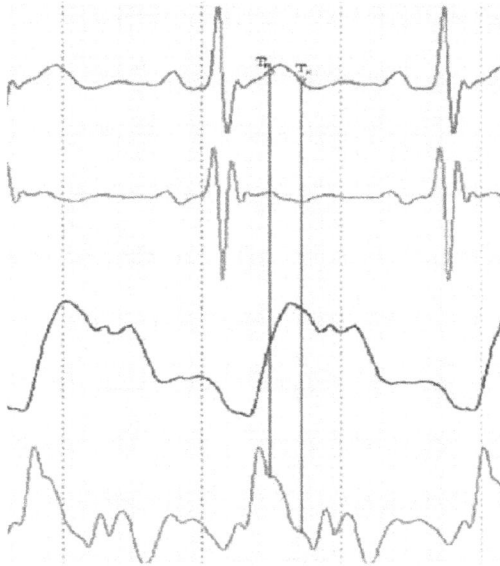

Fig. 73. Maximum of wave T coincides with pressure peak on rheogram

6.2.6. Phase when aortic valve closing begins.

The whole phase of aortic valve closing is a mirror reflection of
phase S–L (sequence of work performed during aortic valve tension) and of phase L–j (aortic valve opening). Actually, the aortic
valve closing phase also consists of two periods. The first period of
this phase begins in point Tк and ends in the dicrotic cavity shown
on rheogram. These processes are practically not evident on ECG
trace. But they provide significant information on rheogram. First
of all, the aortic pressure starts to drop during this period. In case of
normal physiology, this process can be recorded very clearly. In case
of the hindered venous circulation, this process is represented as the
level segment of the rheogram.

189

6.2.7. Phase of aortic valve full closing

The aortic valve full closing phase begins at the instant when the minimum is recorded on rheogram (we mean the minimum that corresponds to the dicrotic cavity) and ends at the instant when Uн is recorded on ECG (Fig. 74). The given phase can not be registered through taking only ECG. Wave U is shown only when critical situation approaches, i.e. when the myocardium is practically at the end of its resources.

It should be noted that the aortic valve closing mechanism is affected by two factors. One of the factors is the pressure drop in the ventricle. Due to the pressure drop, the suction effect occurs and leads to the valve closing. Another factor is the distribution of the pressure that is built up due to constant blood volume in aorta. Both factors prevent the blood backflow quite efficiently.

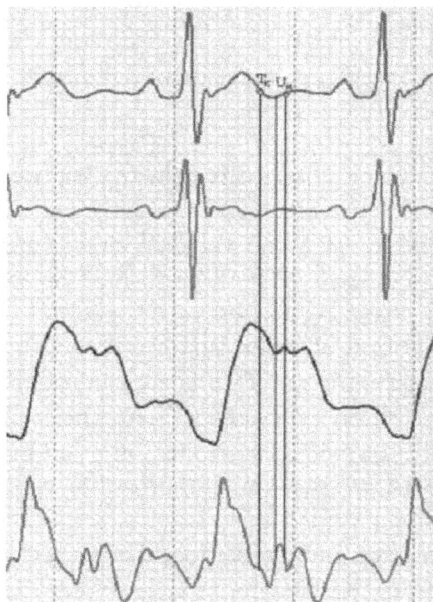

Fig. 74. Point $T_к$ is initial point of phase which ends in dicrotic cavity of the rheogram

6.2.8. Passive venous blood flow phase U_H –P_H

After the aortic valve is fully closed, the passive venous blood flow phase begins (Fig. 75). During this period, the blood flows passively only due to the pressure differential between the center and periphery. In case the vessels elasticity is normal, the pressure drop can be observed on the rheogram. In case the vessels elasticity is insufficient, the straight line is represented on rheogram to indicate the constant pressure, or even the pressure rise can be shown. In greater detail, this process will be described in Section devoted to the diagnosing procedures.

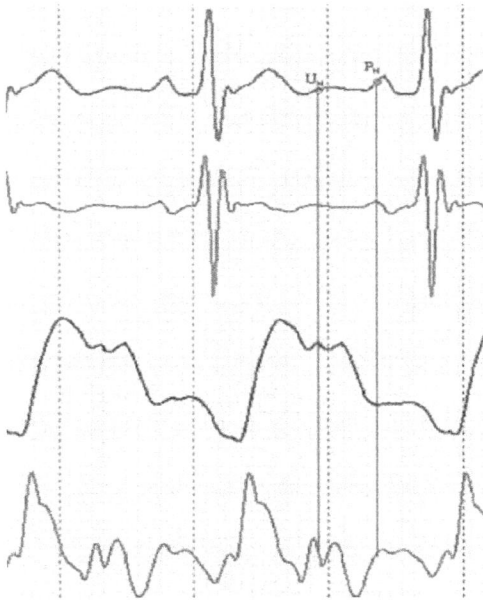

Fig. 75. Passive venous blood flow phase U_H –P_H

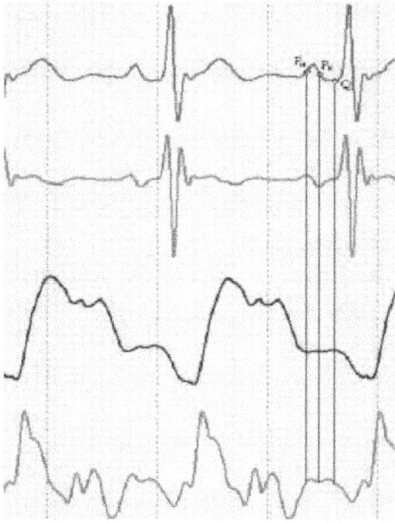

Fig. 76. Atriums and ventricles filling phase P_H–Q

6.2.9. Atriums and ventricles fi[ll]ing phase P_H–Q

This phase differs from other phase[s] the fact that it ensures blood circ[ula]tion in the blood circulatory syste[m] any state of the vessels and myocard[ium] Wave P provides the atriums and [ven]tricles activity, as well as their activi[ty] function of the pump. The basic fea[ture] of this phase is the fact that only [as] this phase is completed, the aortic [pres]sure begins to drop sharply down to [the] diastolic pressure value (Fig. 76).

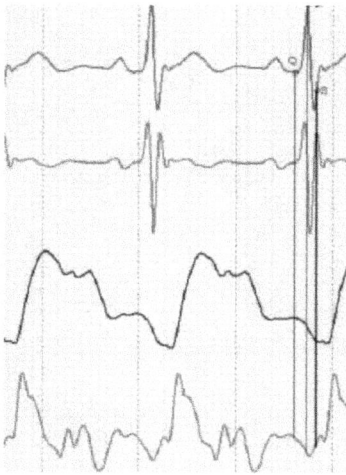

Fig. 77. Phase of pressure build-up in ventricles Q–S

6.2.10. Phase when pressure is built [in] ventricles

During this phase, the pressure de[cay] always observed on the rheogram. [There] is no one case, even the serious path[olog]ical case, when any abnormalities o[f this] phase were detected. Only in cas[e the] aortic valve is weak, the pressure ris[e] [can] occur in the middle of this phase. [How]ever, the pressure decay is always c[learly] seen on the rheogram in phase Q–S [(Fig.] 77).

7. Diagnostics through synchronous recording of ECG trace and rheogram

7.1. ECG and rheogram interrelations used for diagnostics purposes

ECG tracing provides only quantitative characteristics of the hemodynamic processes. The single-channel method employed in CARDIOCODE-1 is highly efficient and enables to obtain reliable characteristics. However, the follow-up investigation indicated that for phase analysis it would be reasonable to amplify the ECG trace with rheogram of ascending aorta being recorded using the same electrodes as for ECG recording. This method allows recording of two signals of different nature through one point. The first signal is an electrical potential used for ECG trace recording. The second signal is mechanical oscillations of blood flow reflected by rheogram. The investigations demonstrated that there are some specific features in ECG and rheogram interrelation (Fig. 78). The essence of these specific features is as follows: there is an order of priority in determining factors redistribution between ECG phases and rheogram. So, PQRSL complex is a determining factor for building up the arterial pressure difference between aorta and vessels periphery. Then, LjTgen complex determines the period for shaping the blood flow pattern. In this case the leading role is played by aorta receptors. After the pattern of blood volume of systole is shaped, the receptors generate the electrical signal to initiate the work of pump mechanism of aorta. The pump mechanism phase corresponds to the length of wave T, namely $T_{н}-T_{к}$.

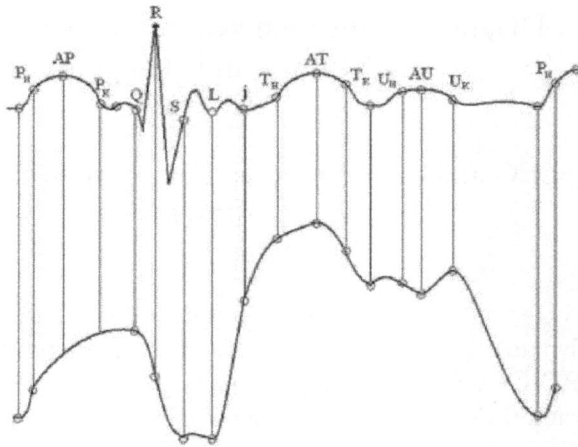

Fig.78. Main points of correlation between ECG and rheogram phases to be in the focus of attention for investigation and diagnostics using CARDIOCODE-2

Then the period follows, within which the systolic blood volume must spread within the entire peripheral blood circulatory system. Irrespective of this period progress and result, the ECG potential (wave P) comes into action.

Thus, since we know how correlate the ECG and rheocardiography phases we can obtain the information on how the normal quantitative hemodynamic parameters are reached, or we can distinguish the source of pathology indicated by certain symptoms.

Let us consider in more detail the informative correlation between ECG trace and rheogram. Figure 78 shows the phases to be considered by the doctor after the quantitative characteristics of the hemodynam-ic processes have been evaluated.

The diagnostics philosophy described in this Section constitutes the basis for CARDIOCODE-2 software that is described in section "Diagnosis made by doctor". While considering the phases in sequence provided by this Section, the doctor has only to choose the necessary information and click the appropriate button on display.

194

The selected data will be recorded in diagnosis sheet. Specific cases are to be registered by the doctor using the computer keyboard.

7.2. Cardiovascular system functions which can be assessed using CARDIOCODE-2

7.2.1. Assessment of aorta valve function

The examination should begin with point S. This is a key point in heart and vessels work mechanism. In the interval preceding this point the pressure in ventricles is built up. At the instant when the point is fixed, the aorta valve opening is initiated. The valve opening has two phases. First phase, S–L, consists in valve tension. The second phase, L–j, consists in valve opening.

In the beginning, the segment preceding point S should be assessed visually. Normally, the pressure in aorta must not rise before point S is reached, since the valve must be closed. If the valve is weak, i.e. it has some pathology, then during myocardium contraction the blood can be supplied t o aor t a th at w i l l b e ma n i fe sted by pre s su r e build-up which is indicated on rheogram. This variant is shown in figure 79. The result will be a raised pressure in aorta.

When valve is in normal condition, the pressure in aorta must not change in the interval between points S and L. The valve must open only after passing point L and pressure in aorta begins to rise rapidly. This process takes place till point j is reached when the valve opening is over. The instant of valve full opening is indicated by local maximum on rheogram first-order derivative curve. If pathology is more pronounced the pressure build-up can be observed beginning from point R and even earlier.

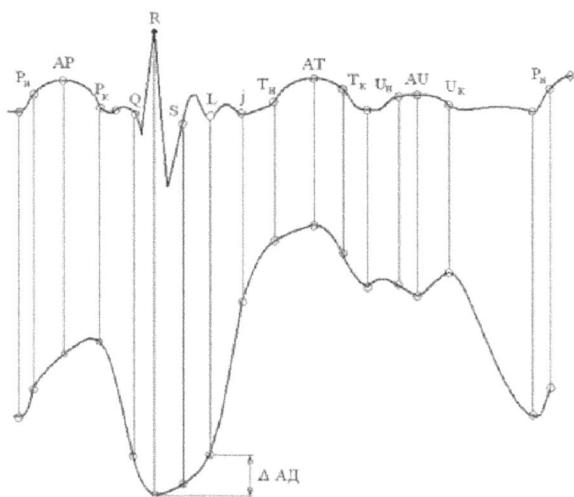

Fig. 79. Valve pathology resulting in pressure raise in aorta within phase R–L
where the pressure must be constant

Disfunction can occur due to two reasons. The first one is anatomical changes in valve shape. Raised pressure in lesser circulation circuit can be the second reason.

Figures 80, 81 show diagrams obtained for a weak valve and recorded through CARDIOCODE-2.

The rheogram shows that the pressure increase is available in phase QR. This is a criterion of significant weakness of aorta valve function.

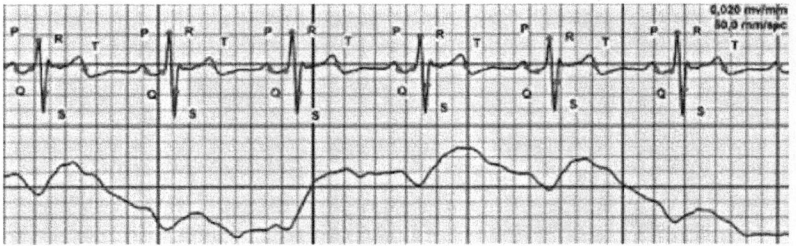

Fig. 80a. Pressure rise in aorta must begin after point L is passed. It is obvious from rheogram that pressure begins to rise in phase QR. This testifies to considerable weakness of aorta valve function

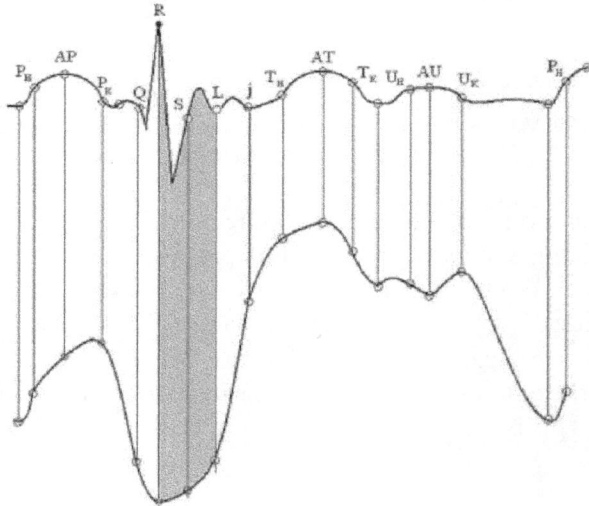

Fig. 80b. Animated figure used for diagnostics of "considerable weakness of aorta valve"

In CARDIOCODE-2 manual, Section "Diagnosis made by doctor", the prompting information is provided in the shape of animated drawing. To register the diagnosis it is enough only to select the shape of the drawing corresponding to diagrams which have been recorded and press the button for getting the diagnosis. The animated prompt is shown in Fig. 80, b.

As a rule, the diastolic pressure of a patient whose diagnosis is similar to above-mentioned diagnosis, is 95 to 100 mm Hg. At emotional tension the pressure can exceed 100 mm Hg.

Fig. 81 shows the case when pressure begins to rise in phase R–S. At this the measured diastolic pressure is as a rule 80 to 95 mm Hg. This case can be qualified as considerable weakness of aorta function because the hypertonia is already in progress though it is an initial stage for the time being.

Valve function can be weakened inconsiderably. Very often the rise of pressure is observed within the phase of the valve tension S–j. This is shown in Fig. 82a. Figure 82b shows the animated prompt provided by CARDIOCODE-2.

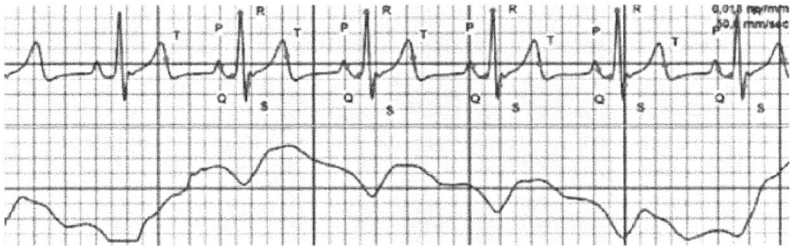

Fig. 81. Rise of pressure in aorta begins in phase RS closer to point S

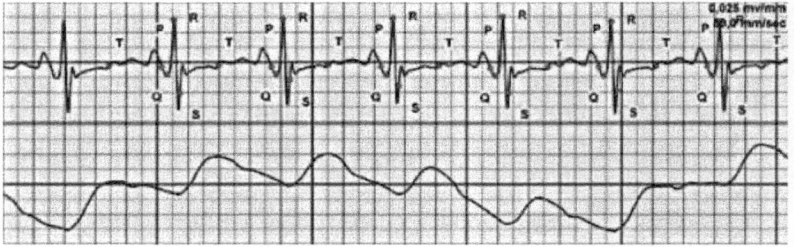

Fig. 82a. Opening of aorta valve in the valve tension phase S–j

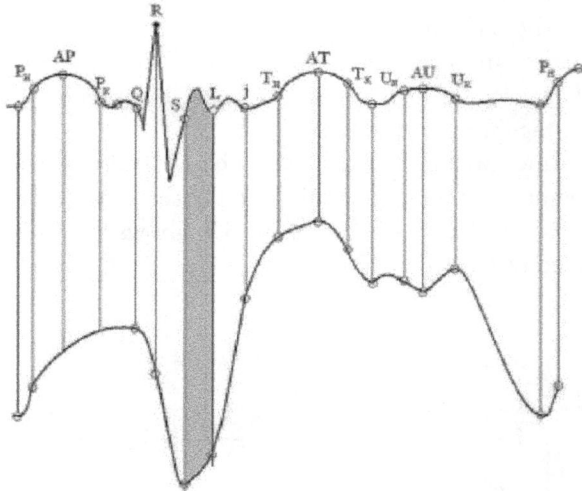

Fig. 82b. Animated prompt for "aorta valve function is weakened"

Sometimes, the delay in aorta valve opening is registered. The valve opens in phase j–Tgen. It's a rare case and for the time being it can not be labeled as a pathology condition since it is not accompanied with definite symptoms (Fig. 83a). In this case only slight difference between measured systolic and diastolic pressures is observed. But this delay may not be observed during the orthostatic test when changing over from horizontal to vertical position.

7.2.2. Anatomical features of aorta valve

The aorta valve anatomical features which are not labeled as pathological features are observed very seldom. They can be referred to as individual anatomical features. With their presence a patient can feel quite well. But standard ECG trace provides the information about severe myocardial infarction. Figures 84 and 85 provide illustrations of such ECG trace recording.

Both male-patients have ECG traces during their lives. None of the doctors managed to make a precise diagnosis. However, the arrythmia occurs on exertion of both patients.

Under close examination it can be seen, that, in spite of evident peculiar features of S–L–j complex, the aorta valve function is normal. Arterial pressure parameters are also within normal range.

7.2.3. Aorta elasticity

Aorta elasticity is determined in phase j–T$_{gen}$. The processes shown in figures 86, 87, 88 reflect the stroke volume distribution within the aorta depending on aorta elasticity parameters. Three criteria are selected for diagnostics: norm, slightly lower than normal and essentially lower than normal. Figure 86 shows the case of normal elasticity of aorta. Coincidence of rheogram tops and waves T on ECG trace is a main criterion of normal elasticity of aorta. At this, the arterial pressure curvatures of rise and decrease are in symmetry relative to rheogram tops.

Decrease in aorta elasticity (slightly lower than normal) is expressed in increase of time required for stroke volume distribution along the aorta length. At this, phase $j-T_H$ is extended. These processes are described in detail in Subsection 6.2.4. Figure 87a shows the case of the elasticity is slightly lower than normal elasticity. Flat top of rheogram extended beyond point Tк is a main criterion of this case. Figure 87b shows the animated prompt for the case of aorta elasticity slightly lower than normal. The prompt is used in Section "Diagnosis made by doctor" of CAR-DIOCODE-2 software.

If aorta elasticity is essentially lower than normal, the pressure rise is negligible after the aorta valve opens fully (point J). On the rheogram this will be shown by the inflection of arterial pressure wave leading edge beginning in point j. The pressure decrease shown on rheogram begins behind wave T on ECG trace. Figure 88b shows the animated prompt for the case of aorta elasticity essentially lower than normal.

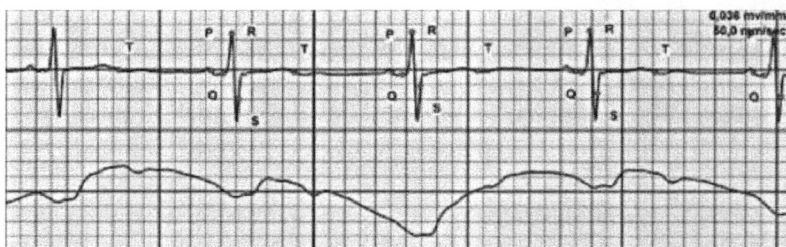

Fig. 83a. Delay in aorta valve opening. Rise of pressure is observed past point

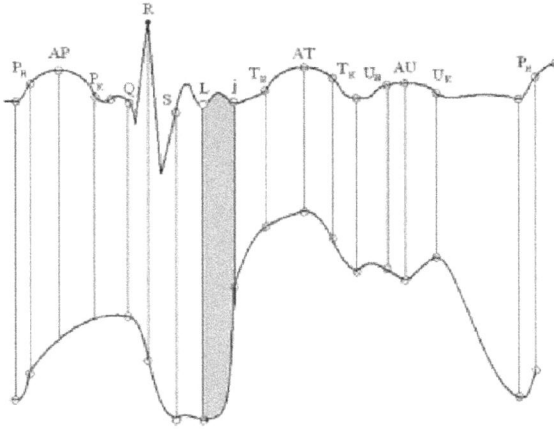

Fig. 83b. Animated prompting for "valve opening with delay"

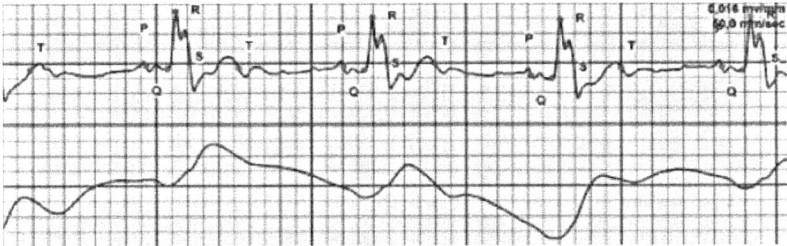

Fig. 84. ECG trace of a male-patient of 60 years old. It is more reasonable to refer this ECG trace to individual anatomical features

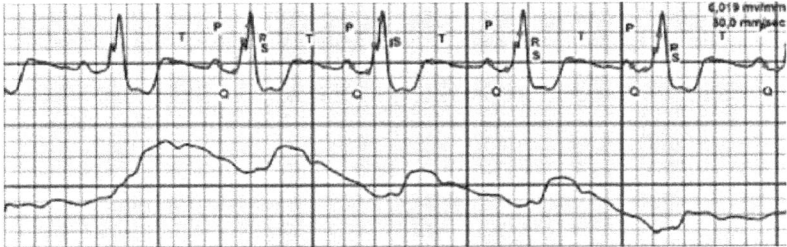

Fig. 85. ECG trace of a male-patient of 59 years old similar to ECG trace shown in Fig. 84

203

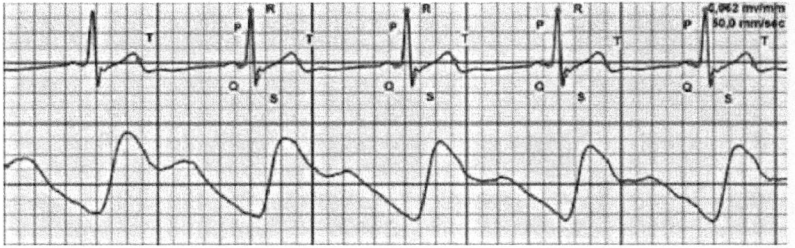

Fig. 86. Normal elasticity of aorta. Top of rheogram coincide with top of wave T

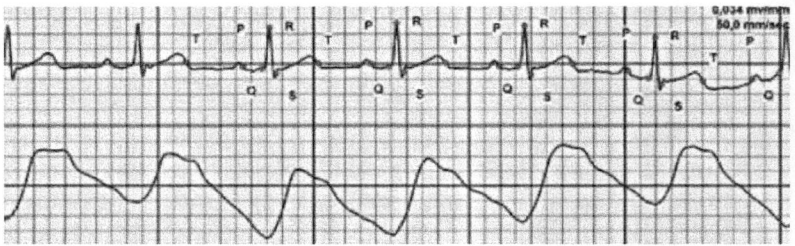

Fig. 87a. Aorta elasticity is slightly lower than normal. Flat top of rheogram is a main criterion

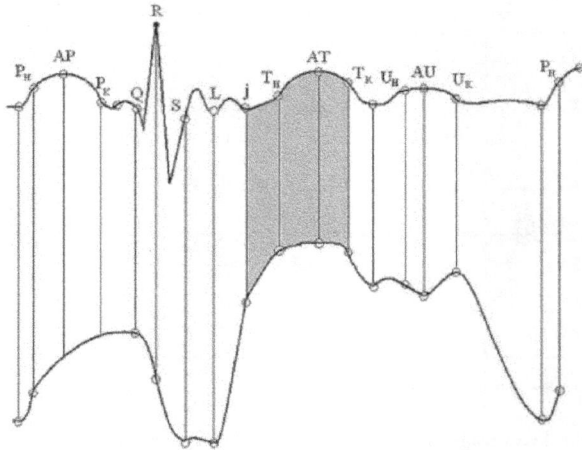

Fig. 87b. Animated prompt for the case of aorta elasticity slightly lower than normal

204

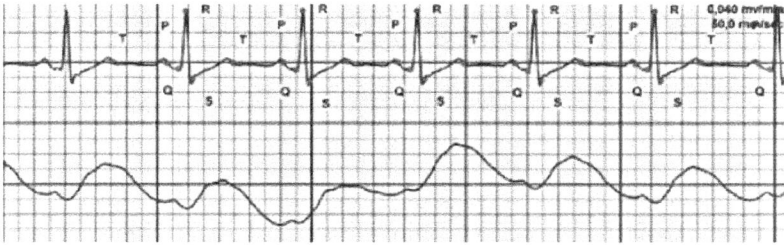

Fig. 88a. Aorta elasticity is essentially lower than normal. Inflection of the wave leading edge on rheogram beginning in point j is a main criterion

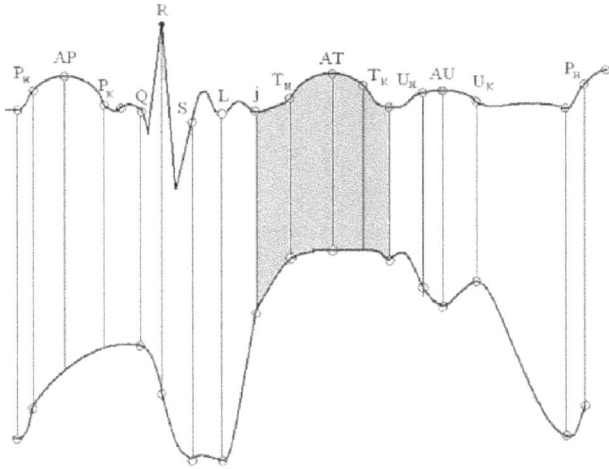

Fig. 88b. Animated prompt for the case of aorta elasticity essentially lower than normal

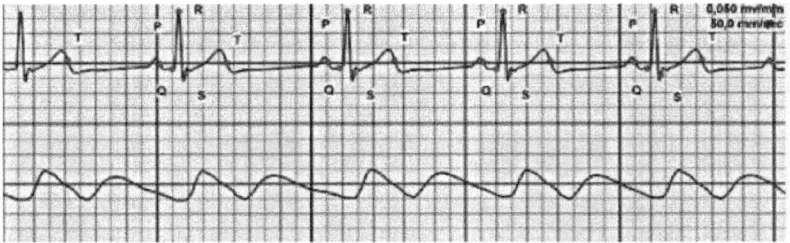

Fig. 89. Rheogram shows the diagnostic criterion of aorta flaccidity (aneurysm)

One more state of aorta elasticity is possible, that is the aorta dilation. In some cases this phenomenon is diagnosed as aortic aneurysm. The orthostatic tests have shown that the aorta volume can change in various positions of the body depending on aorta elasticity. This never occurs in case of atherosclerosis because the aortic walls become inflexible. Excessive dilatability of aorta (aorta flaccidity) is opposite to atherosclerosis state. This term is more suitable for description of phenomena recorded during orthostatic test than the term "aortic aneurysm" (aorta dilation). The term "aorta dilation" implies rather a static condition than dynamism of pathological process. From our point of view, the aortic aneurysm can be diagnosed if the diagnostic criterion is revealed during orthostatic test in both cases. If it is revealed only in one case then we can diagnose the aorta flaccidity.

Figure 89 shows the rheogram for the case of aorta flaccidity. It is evident from the figure, that beginning from the instant of valve opening in phase $j-T_{_\text{H}}$ the pressure does not rise, but decreases to diastolic pressure. After the aorta valve is closed, the pressure is compensated to the level at which it began to decrease in phase $j-T_{_\text{H}}$. The blood backflow to ventricle does not occur because the ventricle functions normally. From the physical point of view the process which determinates the rheogram shape is described in Subsection 8.1. In present section we only state that excess aortic walls stretching (due aorta widening caused by stroke volume output) increases the aorta volume thus decreasing the arterial pressure.

206

The aorta flaccidity is manifested as follows: a patient feels as if he (she) lacks air to breath and needs to come up to the open window to draw a deep breath. This can be accompanied with vertigo and atonia.

7.2.4. Contractile function of interven-tricular septum

Condition of contractile function of in-terventricular septum is determined by ECG trace only. The rheogram does not record parameters for this condition. The compensatory function exists to apportion contraction between interventricular septum and myocardium thus enabling to maintain the normal pressure in aorta. Assessment of contractile function makes possible to estimate the efforts of the body to maintain the normal arterial pressure. Figure 90a shows normal ECG trace for interventricular septum contraction phase Q –R. In normal condition, the absolute values of waves R and S amplitudes must be equal. Figure 90b shows animated prompt for normal condition of this phase.

Subnormal contracticle function of in-terventricular septum is indicated by decrease of wave R amplitude in proportion 1:2 as compared with wave S amplitude (Fig. 91a). Figure 91b shows the animated prompt for diagnosis "Contractile function is lower than normal" for the case under consideration.

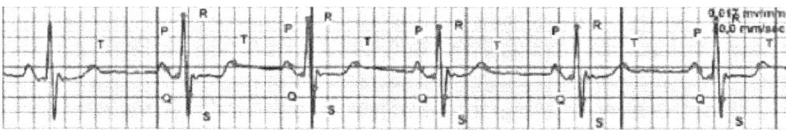

Fig. 90a. Normal amplitude of wave R which testifies to normal condition of interventricular septum contractile function

207

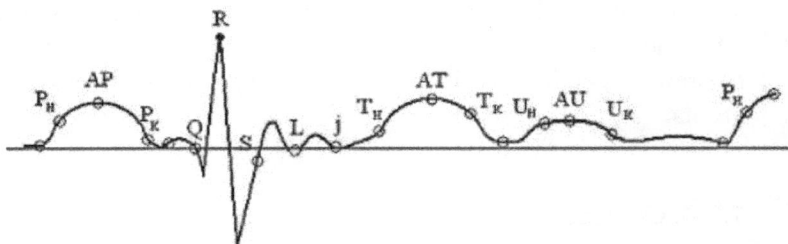

Fig. 90b. Animated prompt for diagnosis "Norm" confirming the normal condition of interventricular septum contractile function

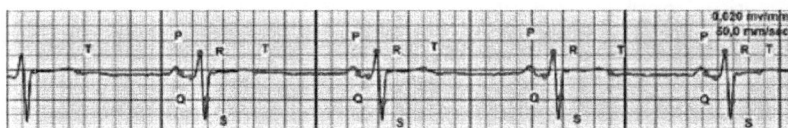

Fig. 91a. Subnormal contracticle function of interventricular septum indicated by decrease of wave R amplitude

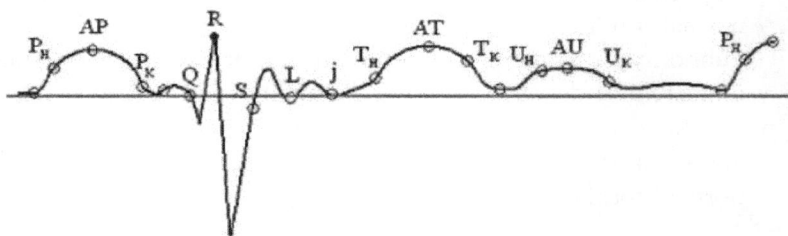

Fig. 91b. Animated prompt for diagnosis "Contractile function of interventricular septum is lower than normal"

For the extreme case when wave R amplitude is lower than wave T amplitude, the diagnosis "Essentially lower than normal" is made (Fig. 92a). At that, the he-modynamic parameters may be within the norm, but the price for this condition is too high for the body. The patient will suffer from exercise load. Figure 96b shows the prompt provided by CARDIOCODE-2 software for this diagnosis.

During orthostatic test, when cell membranes of interventricular septum are subjected to swelling (Subsection 8.1), the abrupt decrease of wave R amplitude (down to isoline level) can be recorded (Subsection 5.4).

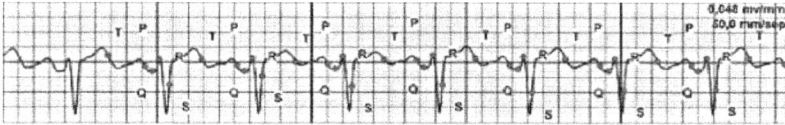

Fig. 92a. Considerable diminishing of function of interventricular septum. Wave R is very small.

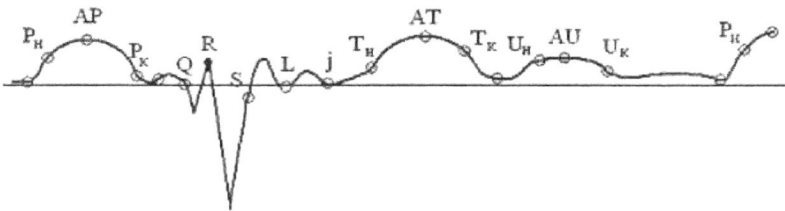

Fig. 92b. Animated prompt for diagnosis "Contractile function of interventricular septum is essentially lower than normal"

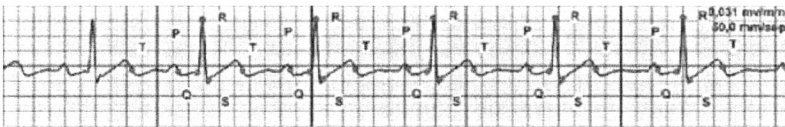

Fig. 93. Essential decrease of contractile function of myocardium. Actually, wave S is absent

7.2.5. Contractile function of myocardium

Most probably the subnormal contractile function of myocardium reflects the capability of a person to carry out the extra physical exercise. During orthostatic test the amplitude of wave S does not change whatever myocardium malfunction would occur. Similar to interventricular septum, the myocardium contractile function diagnosis can be as follows: normal, lower than normal, essentially lower than normal. The prompts are also similar, the only difference is that the prompts show the reduction in length of wave S. Figure 93 shows the essential decrease of contractile function of myocardium.

7.2.6. Compensatory interrelation between contractile function of interven-tricular septum and myocardium

This portion of the diagnosis is introduced to emphasize the importance of Q–R–S complex functioning. Very often it is necessary to know about reliability of heart work when dealing with pathological condition of other portions of cardiovascular system which require surgical or some other treatment. If absolute values of waves R and S amplitudes are equal, then the heart has the optimal reserve to withstand high physical and emotional stress. If amplitude of one wave dominates over the other, then it is necessary to be very careful and analyze the possible medical treatment many times.

The animated prompts are provided by the instrument software to assist the doctor in diagnostics process. The prompts are similar to those described in Subsections 7.2.4 and 7.2.5.

7.2.7. Venous blood flow state

After assessment of heart and aorta functioning we can proceed to diagnostics of peripheral vessels and venous blood flow. The venous blood flow is assessed by the slope of the rheogram curve segment corresponding to ECG trace segment located to the right from wave T. In normal condition this part of rheogram must be angled to 45°. In this phase the blood flow is natural. The aorta valve is already closed. If the heart does not cope with blood flow pattern formation (Subsection 2.2), then wave U appears on ECG trace and enhances the pumping function of the heart. At this, no special alternations would appear on the rheogram in this phase.

The rheogram slope characterizes the blood flow. If it is hindered then the flow is smaller and the slope is closer to horizontal line. If it is restricted by considerable resistance of peripheral vessels and veins, then the pressure rise is shown on the rhe-ogram. This pressure rise can exceed the systolic pressure. In any case, the pumping function of the atriums facilitates the blood moving to atriums beginning from the instant of wave P occurence on ECG trace. The blood circulation cycle is closed.

Figure 94 provides a typical example of pressure rise in the phase due to increase of resistance to blood flow. The cause of increase of resistance can not be found through this method. A number of factors can cause this condition.

Very often after the insult, pressure stagnation can be observed in the phase under consideration. Pressure stagnation can be also caused by pathological condition of venous system.

Pressure stagnation is shown in Fig. 95a,b.

The appropriate animated prompts are provided in the instrument software. They are shown in Fig. 96a,b,c.

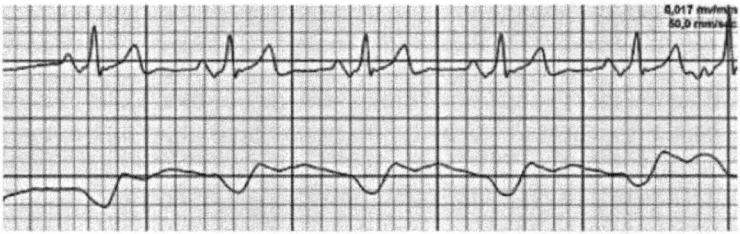

a) Equal pressures

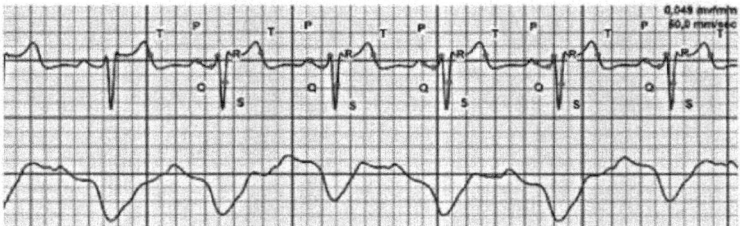

b). Pressure is higher than systolic

Fig. 94. Flow rate in phase of passive venous blood flow decreases due to increase of resistance to venous blood flow. Both pressure equality (a) and pressure rise beyond systolic (b) can be observed

7.2.8. Function of lungs

Pulmonology investigations revealed one more diagnostic criterion, this is the lungs condition. It is not related with cardiovascular system directly but affects it indirectly. This criterion is based on directly-proportional relationship between pulmonary function and the level of rheo-gram isoline modulation by respiratory undulation. If pulmonary function is normal, the rheogram is recorded along the isoline. If function is lower than normal, then respiratory undulation modulates the isoline. Figures 95 and 96 show the cases of pulmonary function conditions lower than normal and essentially lower than normal. It can be seen that respiratory undulation not only

212

modulates the rheogram isoline, but (in case of pulmonary function essentially lower than normal) it also modulates the amplitude of the rheogram.

A conclusion can be made that lungs perform the function of a buffer: when in normal condition, they prohibit the pulmonary function effect on hemodynamic parameters. However, when the pulmonary function is lower than normal (to be more precise, when the structural changes occur resulting in partial loss of function) the respiratory undulations begin to affect the rheogram recording for ascending aorta. To avoid this effect, it is recommended to a patient to hold his breath during rheo-cardiography procedure.

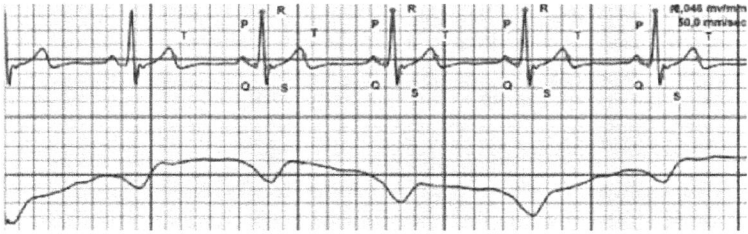

a)

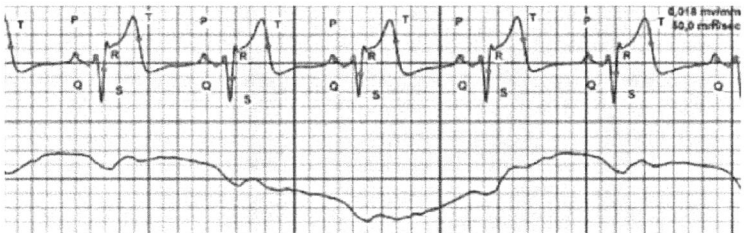

b)

Fig. 95 a,b. Pressure stagnation

213

7.2.9. Preinsult condition

It was noticed, that in some cases the rhe-ograms have distortions which look like peaked waves similar to Q–R–S complex. Distortions are of local character, and differ in amplitude. At more scrupulous investigation it is evident that these impulses are generated in initial and end points of the aorta valve opening or closing. Later on it was found out that impulses have regular character and do not bear distortions in rheogram. These impulses bear the early warning about probability of insult. They are not considered as diagnostic criteria for preinsult state of the body.

Figure 99 shows these impulses. It is characteristic that these impulses show up only on the rheogram, while the ECG trace remains unchanged. Only in some cases Q–R–S complex shape is changed and wave amplitude in valve opening phase S–L–j is increased essentially.

In Fig. 99 it can be seen that impulses have negative and positive polarity. Abrupt changes of pressure are not characteristic for normal hemodynamics of cardiovascular system. The occurrence of the impulses is not accompanied with any external manifestations.

The symptoms appear when the amplitude of impulses is increased and when impulses change over their peaked shape to rectangular shape (Fig. 100). At this, the typical occipital pain and brachial muscle obdormition occur.

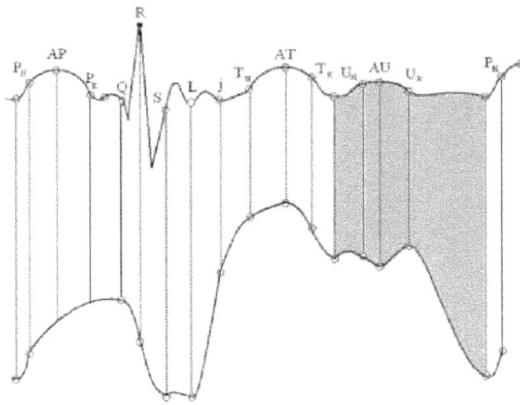

a)

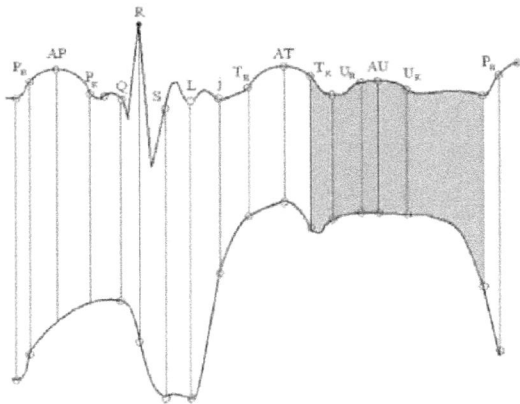

~)

215

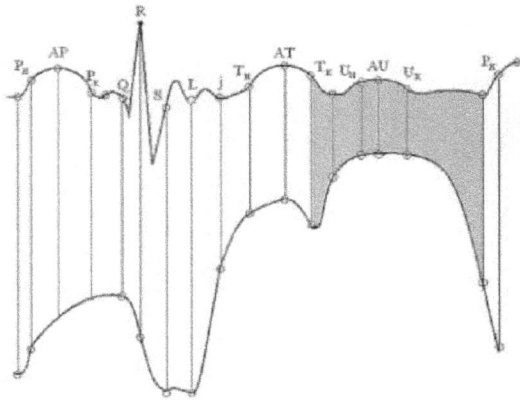

Fig. 96. Animated prompts used to diagnose the condition of venous blood flow:
a) normal; b) hindered; c) essentially hindered

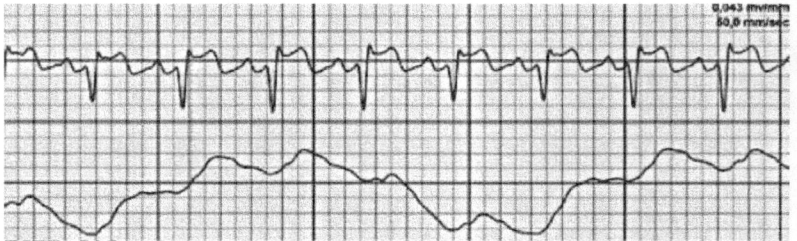

Fig. 97. Pulmonary function is lower than normal

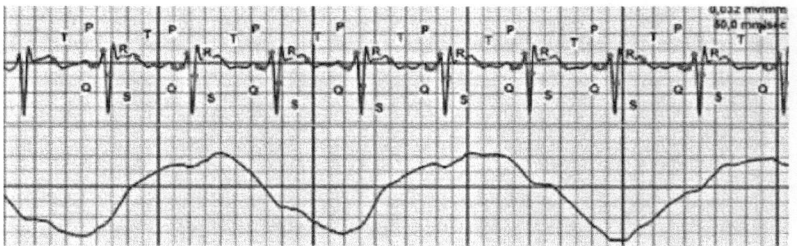

Fig. 98. Pulmonary function is essentially lower than normal

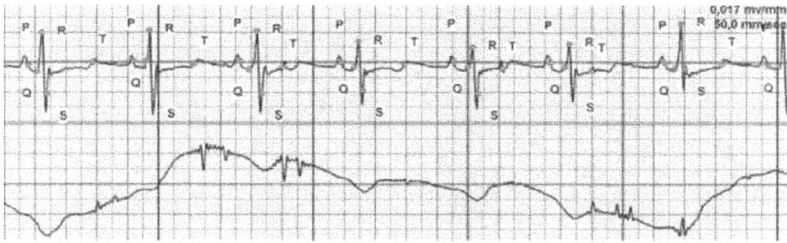
Fig. 99. Impulses of positive and negative polarity

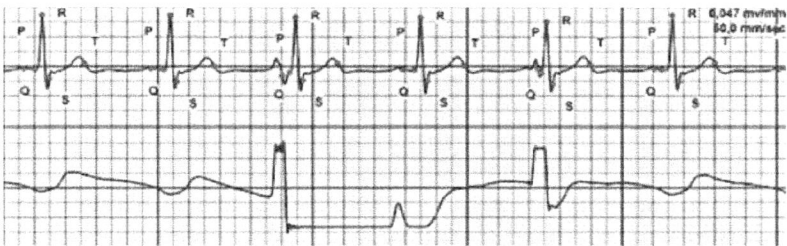
Fig. 100. Typical pressure impulses of rectangular shape accompanied with severe occipital pains and deterioration of visual acuity. The male-patient had a phobia, but he consulted a doctor and avoided the insult

Impulses can show up during or tho st atictest when the position of body is changed from horizontal to vertical. At these no symptoms of preinsult condition are shown.

The preinsult state of some patients was cured in ambulatory condition taking into consideration the symptoms and using the methods for pressure decreasing. Ex-sanguination was made in location of seventh cervical vertebra with use of a special-purpose acupuncture needle hammer, then a vacuum glass-cup was applied. The required therapeutic effect was obtained.

Impulses can have the negative polarity. In this case the patient suffers from vertigo. But these cases are few and far between.

The authors designated the recorded pressure impulses as hemoimpulses.

217

The Section "Diagnosis made by doctor" provides the similar animated prompts. The doctor has only to choose the appropriate prompt and state the diagnosis.

8. Updating of cardiovascular system pathologic behavior theory based on cardiac cycle phases analysis data

Data obtained in the course of investigation made with use of CARDIOCODE necessitate reviewing of existing theories related to cardiovascular system pathology progress. It was revealed that the theories had a number of blind spots and ambiguous definitions of biophysical processes taking place during appropriate cardiac cycle phase. It is true since the phase analysis method set forward by the authors enables to study the dynamics of heart and vessels more scrupulously. It was infeasible so far.

This section provides the systematized data obtained during clinical investigations of cardiac cycle phase processes. The data include main clarifications of theoretical concept of cardiovascular system pathological behavior.

The pathologies may progress in two ways.

First way: slow temporal progress i.e. thrombosis.

Second way: rapid temporal progress. i.e. myocardial infarction.

Common initial conditions: disorder of muscle cells energetic characteristics due to malfunction of intermembrane transport of biochemical elements.

Now let us consider the pathology progress in more detail.

8.1. Common initial conditions for pathology progress. Malfunction of intermembrane transport of biochemical elements.

A living cell membrane is a main element to maintain biochemical reactions during muscular activity[144] [64]. The membrane structure includes complex receptor sites that ensure various cell activity[145] [64]. These sites incorporate molecular structures the main component of which is carbon[146] [64]. The molecular structures support the work of calcium pump, and the contractile function of muscles and normal course of oxidizing processes in cells depend on this pump. The membrane molecular structure disorder is conditioned by various factors. The following are the most permanent factors that induce the membrane pathology: lack of carbon determined by level of carbon dioxide in blood, presence of toxic substances and excess amount of free radicals in body. Each factor contributes to disorder in cell internal/external environment feedback that results in disbalance of energy production in cells[147],[148] [65, 66].

As a consequence, at initial stage, the vessels elasticity is decreased that can be manifested locally or at system level[149] [67]. This results in changing of blood circulation parameters, first of all the arterial pressure parameters[150] [68]. But the arterial pressure parameter is a

144   Leninger A. Fundamentals of Biochemistry, 3 Volumes. Translation from English into Russian. M., "Mir", 1985, 1056 pages.
145   In the same reference. - P. 45-46.
146   In the same reference. P.47.
147   Kapelko V.I. Disturbance in Energy Generation in Cardiac Muscle Cells: Causes and Consequences // Soros' Education Journal. Biology. 2000. Vol.6 No.5 Pages 14-20 Access: www.pereplet.ru/obr azovanie /stsoros /99 2.html
148   Kapelko V.I. Creatine Phosphokinase Way of Energy Transport in Sarcous Cells // Soros' Education Journal. Biology. 2000 Vol.6 No.11   Pages 8-12.
149   Rubtsov A.M. The Role of Sarcoplasmatic Reticulum in Regulation of Contraction Activity of Muscles // Soros' Education Journal. Biology. 2000. 39. Pages 17-24 Access: www.pereplet.ru/obrazovanie/stsoros/1066.html
150   Kapelko V.I. Blood Circulation Regulation // Soros' Education Journal. Biology. 1999. No.7 Pages 79-84 Access: http://www.pereplet.ru/cgi/soros/readdb. cgi?f=ST750

general integral parameter. It should not be used for process prediction because it only ascertains the actual state. Most informative differentiated diagnosis obtained through the cardiac cycle phase analysis should be used for prediction. The phase analysis enables to detect the local pathological change of vessels elasticity caused by muscle contractile function disorder.

It should be stressed again that vessels elasticity reduction is associated with disturbance in energetic processes in muscle cells. The process develops and leads to in-tracellular edema which results in arteries walls swelling. The process runs gradually and is manifested by personal subjective symptoms.

Let us consider in detail the biophysical processes which characterize the diagnostic criteria of gradual swelling of arteries, heart and vessels walls manifested on ECG trace.

The CARDIOCODE enables the simultaneous recording of ECG trace and rheogram to ensure reliable detection of muscle vessels elasticity changes in interventricular septum, myocardium and in ascending and descending portions of aorta.

According to relationships established in Section 2.5, Q–R–S complex characterizes the phases of interventricular septum and myocardium contraction. The ECG proper represents an electrical potential between two local points. In case of artery walls swelling, the orthostatic test shows the considerable reduction of R wave amplitude. At this, the personal symptoms are changing with the main ones being lack of air to breath and vertigo. Why does ECG trace show the R wave amplitude reduction? The answer can be found in elementary physics. Let us pass the current through flexible tube filled with liquid, with recording electrodes being arranged lengthwise on the tube. When the tube is in horizontal position, the potential correlated with amount of liquid between the electrodes is recorded (Fig. 99a). According to Ohm's law, the potential is equal to:

$$U = R \cdot I, (23)$$

where
R is resistance of liquid between the electrodes;
I is current intensity (which is constant in this case.)

When the tube is set vertically, the amount of liquid between electrodes is larger due to the tube expansion in horizontal plane (Fig. 101b). At direct current the resistance gets lower thus causing the voltage reduction.

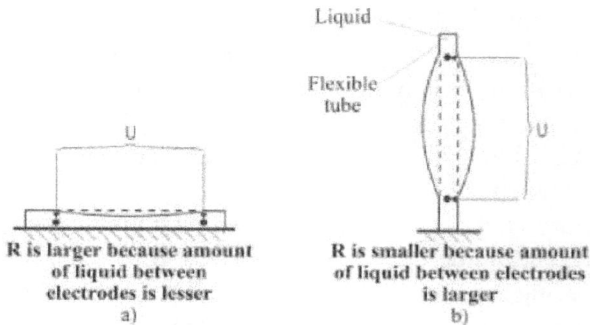

Fig. 101. Changing of potential U depending on tube expansion

The similar processes are recorded during the ECG taking. At this, the described effect of liquid redistribution is observed in pathologic wall of the vessel. During the orthostatic test, the wave R potential does not change if elasticity of vessels in inter-ventricular septum muscles is normal.

The similar biophysical phenomenon can be observed in ascending aorta. In this case, the pathology is shown not on ECG trace, but on the rheogram in the shape of abrupt decrease of pressure in systolic phase while it should grow up.

Let us consider the result of durable swelling of cell membrane. At a stage of swelling progress (namely, when during the orthostatic test the wave R amplitude decreases to isoline drawn from point Q) the pathology can progress in one of the two above-mentioned ways, that will depend on quality of reactions aimed at steady state (homeostasis) maintaining in the body[151] [68] (Fig 102).

Two antipodal systems are existing in the body: blood coagulation system and an-ticoagulation system. Predominance of one system over the other is determined by quality of homeostasis reactions which determine the state of the body and personal symptoms. For example, the blood coagulation system predominants when the body is subjected to some stress and percentage of lymphocytes in blood equals to 5 to 19 while the normal percentage is 20 to 40[152] [68]. The blood coagulation decreases (percentage of lymphocytes is 28 to 33) during physical exercise. During normal activity (percentage of lymphocytes is 28 to 33), equilibrium between two systems is observed. At high activity (percentage of lymphocytes is 30 to 40), the blood coagulation decreases. At hyper-activity, (percentage of lymphocytes is 41 to 65) blood coagulation can be of various level, since this state develops to stress at which the blood coagulation dominates.

So, to prevent lysis (Fig. 102) the body can activate a protective function, namely cholesterol plaques can be shaped. This function is aimed at blocking the oxygen access to cells with swelled membranes. The function can be activated or not depending on current condition of the body. Thrombosis development process is shown in Fig. 103. It is described in more detail in subsection 8.2.

It is important to note, that both ways lead to pathological conditions. Lysis results in infarction, and thrombosis can result in formation of thromb. The only difference is that the second way is longer and critical condition becomes evident much later. That is way it is very important to take effective therapy steps to preclude membrane swelling in case some changes in phase amplitude have been revealed during orthostatic test.

Here we have a paradoxical situation. When the body is in a sound physical condition, the probability of infarction is considerably high.

---

151   Garkavi L.Kh. Activation therapy. Rostov-on-Don. Published by the Rostov State University, 2006. 256 pages.
152   In the same reference.

But when the body is in bad condition, the probability of infarction is much lower. In this case, the long-lasting thrombus formation process is more probable which can also result in catastrophic condition.

Garkavi L.Kh. Activation Therapy. Rostov-on-Don. Published by the Rostov State University, 2006, 256 pages. 102 In the same reference.

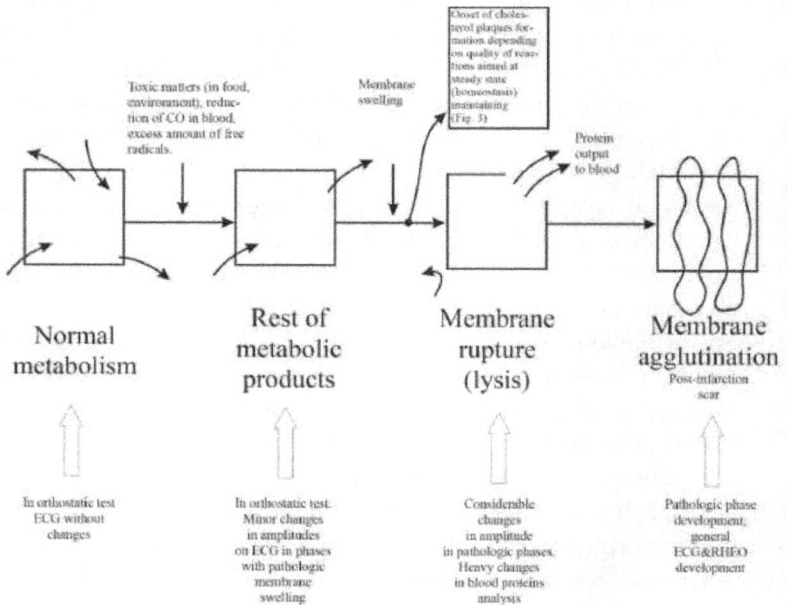

Fig. 102. Cardiovascular system pathology behavior. Pathology assessment criteria.

Cells function
disorder (8.1)
Cholesterol
plaques
formation

Deterioration

Fibrin polymeriza-
tion, increase of con-
centration of throm-
bin coagulation acti-
vators

Normal blood flow at
normal cells function

Formation of plaque as a
protective function prevent-
ing the oxygen and nutrients
supply to damaged cells

Blood flow disorder
resulting in turbu-
lence

Thrombus

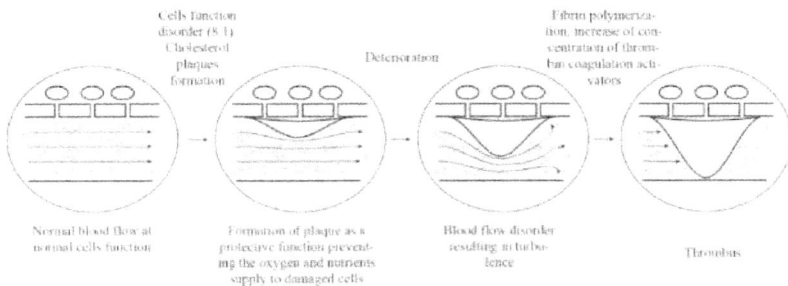

Fig. 103. Thrombosis development through cholesterol plaques formation

At present, investigation of body state using CARDIOCODE is the most effective way for early diagnostics of cell membrane swelling in heart and vessels to take adequate preventive steps described in further sections.

In case the intracellular edema and swelling result in membrane deformation and lysis, the membrane breaks and great amount of enzymes enters the blood flow. Currently, the acute myocardial infarction can be diagnosed based on results of blood serum analysis, first of all aspertataminotransferasa (AcAT) analysis[153] [69].

In fact, we have explored the biophysics of ischemic heart disease development process. The phase analysis ensures a high confidence level for disease localization and degree. Diagnostics criteria obtained through orthostatic test are the most effective.

Section 9 provides the clinical data which enable to systematize the above-mentioned criteria and other criteria revealing the pathological features in myocardium vessels functioning.

153   Yelisseyev Yu.Yu. Coronary Heart Disease (CHD. Stenocardia. Myocardial Infarction. Pathogenesis, Clinical Picture, Diagnostics & Therapy). M., "Kron Press", 2000, 170 pages.

8.2. Thrombosis development

Now we consider the pathological process progression through thrombuses formation. In this case, the formation of a cholesterol plaque is the first stage of the process. The plaque is formed in location of pathological cell, turbulent flow comes to existence downstream of the plaque in distal blood flow. The turbulence launches the mechanism of blood coagulation. In more exact terms, the pattern of blood flow in superfluidity condition (as described in subsection 2.2) is broken.

Cholesterol plaques cause the changes in vessels walls in the form of connective tissue proliferation accompanied by internal walls of vessels impregnation with fat, which results in atherosclerosis. But the atherosclerosis proper does not break the blood flow pattern. The turbulence can occur locally only in case of cholesterol plaques formation[154] [70]. To break the blood flow pattern, the plaque must be of considerable size relative to vessel diameter (Fig. 103).

As distinguished from plaque formation, one more mechanism takes part in formation of thrombuses, that is the distortion of feedback in control of arterial pressure dynamic changes resulting in arterial pressure impulses. These impulses are short with regard to their duration and have high amplitude (Fig. 104). They are described in subsection 7.2.9. The turbulence and distortion are not interconnected, but still they have one common feature: they cause the change in blood flow pattern which leads to occurrence of turbulence.

The cause of abrupt pressure rise is not yet determined positively. There are grounds to consider them to be emerging due to processes in cervical spine distorting the feedback in arterial pressure maintaining at normal level. For example, it was observed that obdor-

---

154   Guriya G.T. et al. Mathematical Model of Activation of Blood Coagulation & Thrombus Growth under Blood Circulation Conditions. A Miscellany of Papers under the title "Computer Simulations & Progress in Health Service", M., "Nauka", 2001, 300 pages.

mition of a hand is a symptom of forthcoming insult. At this, the arterial pressure rises essentially.

Figures 104-107 provide the examples of hemodynamic impulses of various shapes. It is evident that positive and negative impulses exist. Peaked pressure impulses are unhazardous, they only indicate the probability of thrombus formation. Precisely hemoimpulses disturb the blood flow pattern. The progression of flat-topped hemoimpulses is hazardous.

The probability of the vessel break increases where flat-topped impulses are shown. The duration of impulses is of great importance. Sensation of heaviness in the nape is one of the main symptoms in this case.

The existence of negative impulses shows that a patient is prompt to vertigo and sensation of mist. Vascular spasms are very probable. Diagnostics of preinsult state is described in more detail in Section 9.

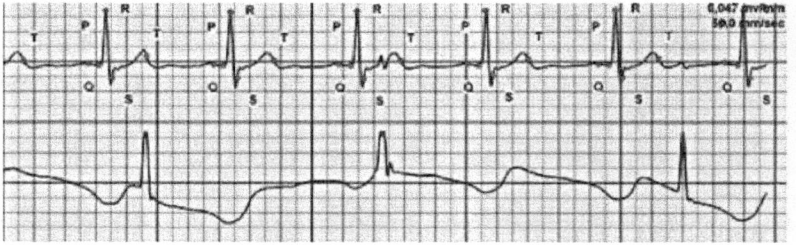

a)

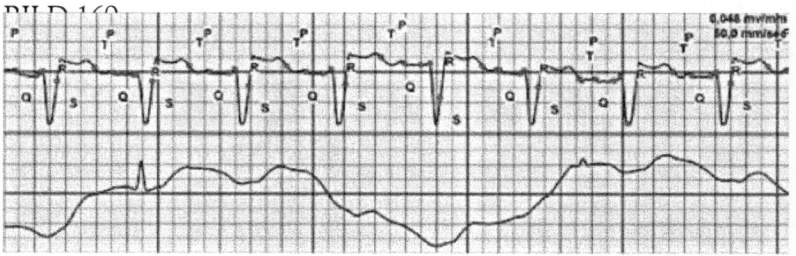

b)

Fig. 104. Peaked hemodynamic impulses of negative amplitude

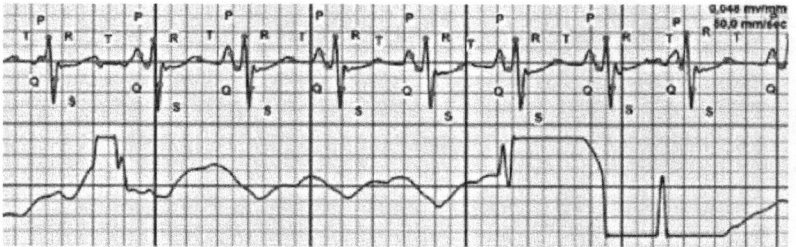

Fig. 105. Peaked hemodynamic impulses of negative and positive amplitude

228

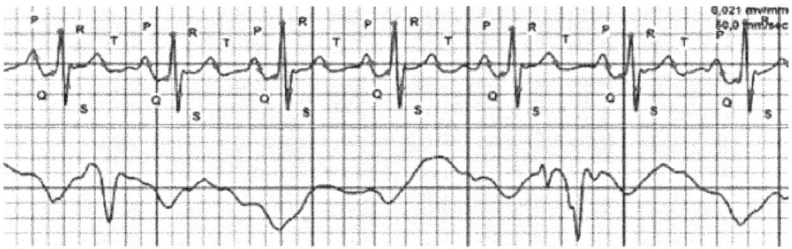

Fig. 106. Peaked hemodynamic impulses of negative amplitude

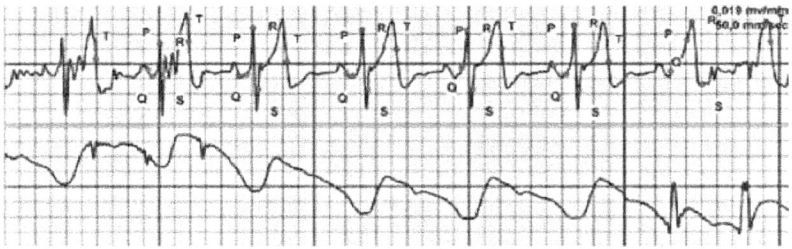

Fig. 107. Peaked hemodynamic impulses of negative and positive amplitude

Part II

Atlas
for functional diagnostics based on cardiac cycle analysis
(ECG trace + rheogram)

9. Practical diagnostics

A detail user manual is always necessary for a new method. This section provides the main criteria to be used for making diagnosis in practice. For better understanding the criteria are considered in the frame of comparative analysis of data obtained through application of CARDIOCODE method and data obtained using the conventional methods which have been widely used in practice. The clinical cases are provided to represent the typical diagnostics challenges in cardiology. Each case is accompanied with comments.

9.1. Relatively normal functioning of cardiovascular system

Figure 108 shows the ECG trace and rheo-gram obtained from middle aged male-patient, who does not go in for sport, keeps to regular life, has no harmful habits, has normal symptoms. The recorded cardi-osignals can be taken for ideal. Note, that it is a rare case in actual medical practice. Let us investigate the sequence and principles of diagnostics using this example.

Any ECG+rheogram record is analyzed beginning from the point where cyclic repetition is observed.

1). In this particular case, we have the practically ideal cyclic repetition. Pulse frequency may vary within fairly narrow limits (it is normal condition), but the periodicity of peak values is very important.

2). It is necessary to pay particular attention to absolute values of waves R and S relative to isoline. "Wave S" is a conventional term, since point S is not located on the wave peak. It is located somewhat to the right in inflection point. Waves R and S amplitudes are equal. If amplitudes were different it would be a signal about degradation of contractile functions of interven-tricular septum and myocardium. Reduction of wave R amplitude is a criterion of inter-ventricular septum contractile function degradation. Reduction of absolute value of wave S is a criterion of myocardium contractile function degradation. In case under consideration the absolute values of waves R and S amplitudes are equal, that is an absolutely normal case.

3). Waves T and P amplitudes are well-defined. Wave U is not shown distinctly. Wave T is twice as lower as compared to level of wave R. This is a good characteristic.

4). Rheogram shows a good cyclic repetition. In each phase, arterial pressure amplitudes are of normal values. Amplitudes interrelation is a criterion of normal values.

Date of investigation – 14.06.2004
Date of birth – 26.06.1971
Family name – M.D.

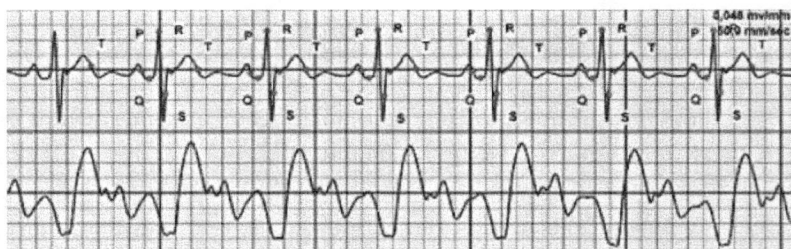

Diagnosis made based on results of phase analysis.
Volume parameters of cardiac and vessels activity are within the normal range.

Diagnosis made by doctor.
Heart and vessels functioning is normal.

SV (ml) – Stroke volume
SV(ml)=55.03 X=0.00   42.01   95.41

MV (l) – Minute volume
MV(l)=4.74 X=0.00   3.62   8.22

PV1 (ml) – Volume of blood in phase of rapid and slow filling of left (aortic) ventricle of heart
PV1(ml)=34.19 X=0.00   24.05   53.85

PV2 (ml) – Volume of blood flowing into left (aortic) ventricle of heart during systole
PV2(ml)=20.84 X=0.00   17.36   41.56

PV3 (ml) – Volume of blood ejected by left (aortic) ventricle of heart during rapid ejection phase
PV3(ml)=32.66 X=0.00   24.91   56.70

PV4 (ml) – Volume of blood ejected by left (aortic) ventricle of heart during slow ejection phase
PV4(ml)=22.37 X=0.00   17.09   38.71

PV5 (ml) – Volume of blood transferred by ascending aorta functioning as peristaltic pump
PV5(ml)=7.53 X=0.00   6.57   10.65

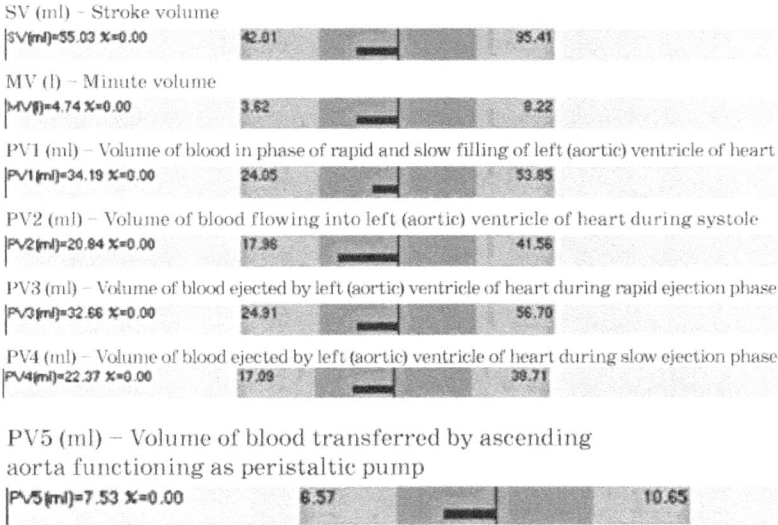

Fig. 108. Standard diagnosis sheet prepared with cardiocode that is issued to a patient

Table 7.

Calculated data of phase characteristics produced with Cardiocode automatically

	QRS	RS	QT	PQ	TT	SV (ml)	MV (ml)	PV1 (ml)	PV2 (ml)	PV3 (ml)	PV4 (ml)	PV5 (ml)
Max	0.080	0.035	0.309	0.072	0.696	42.0	3.6	24.0	18.0	24.9	17.1	6.6
Min	0.110	0.060	0.308	0.072	0.696	95.4	8.2	53.9	41.6	56.7	38.7	10.7
Average	0.098	0.042	0.304	0.051	0.696	55.0	4.7	34.2	20.8	32.7	22.4	7.5
1	0.101	0.042	0.307	0.046	0.686	55.0	4.8	33.5	21.6	32.6	22.4	7.5
2	0.101	0.042	0.307	0.051	0.699	55.0	4.7	34.5	20.5	32.6	22.4	7.5
3	0.101	0.042	0.303	0.051	0.711	54.6	4.6	34.4	20.5	32.4	22.4	7.4
4	0.101	0.042	0.312	0.051	0.728	55.4	4.6	35.3	20.1	32.9	22.5	7.6
5	0.101	0.042	0.315	0.059	0.711	54.6	4.6	34.9	19.7	32.4	22.4	7.5
6	0.088	0.042	0.290	0.059	0.711	54.6	4.6	34.6	20.1	32.6	22.4	7.5
7	0.097	0.042	0.303	0.047	0.669	55.0	4.8	34.3	20.1	32.6	22.4	7.5
8	0.101	0.042	0.312	0.046	0.678	55.4	4.9	33.4	22.0	32.9	22.5	7.6
9	0.093	0.042	0.299	0.059	0.665	55.0	5.0	31.8	23.2	32.6	22.4	7.5

235

5). After general examination of ECG trace and rheogram is over, the doctor can proceed to phase analysis, to be more precise to qualitative assessment of amplitudes within phases. The assessment should begin with analysis of aortic valve functioning, namely it should be started at point designated by S. Segment S–j corresponds to valve opening cycle. Point j is not provided on diagrams. It is omitted for clarity of general information. Doctor should mark point S using the cursor bar. It is important to see time history of arterial pressure changing on the rheo-gram fore and aft of point S. In norm, the arterial pressure on rheogram within phase QRS must not rise before reaching point S. In the segment aft of point S, the arterial pressure diagram must look like a straight line up to the point L (Subsection 6.2.2) where the half-wave comes to an end. These processes are described in greater detail in Section 7 (Fig. 78, point L), while the present Section provides only a flow of thoughts of the doctor when making the diagnosis. In the next phase which corresponds to segment L–j, the valve opens and the stroke volume of blood begins to enter the aorta (Subsection 6.2.3). In this period, the rise of arterial pressure is shown on rheogram. The value of arterial pressure comes up to 1/3 of maximum (systolic) pressure. The following phase is a pause phase j–T□ or it can be referred to as slow ejection phase. At the end of the phase the pressure comes up to 3/4 of systolic pressure. Then the mechanism of pump function of aorta is actuated, and pressure increases slightly to reach its maximum value.

This corresponds to normal correlation of values on ECG trace and rheogram. In real situation we never have come across two patients who have similar recording. However, this fact never complicates the diagnostics. The present description and experience of a doctor enable to analyze the case within a few seconds.

6). The further procedures related to diagnostics are to be aimed at detection of specific deviation from the norm. This record can be used as a reference record.

7). Let us mention some factors to be noted. There is a mechanism of the heat functions self-regulation, which is reflected in hemodynamic parameters observed in phase analysis. The body works to maintain these parameters within the normal condition. This indicates the existence of hemodynamic code similar to genetic code of DNA. (It is this fact that motivated the choice of the name for the method and subsequently the name of the company CARDIO-CODE).

The measured values of 3D hemodynam-ic parameters are represented on page ANALYSIS of the instrument software.

In case of parameters deviation from the norm, the changes in cardiac phases should be studied paying particular attention to relevant segments of ECG trace and rheo-gram. If more than 30% deviation from norm is detected, the patient is subject to hospital treatment.

8). If some symptoms of disease are manifested while the hemodynamic parameters of the patient are normal, the doctor should carefully investigate the local painful zones on the chest and analyze the level of health in general. The symptoms and phase changes on ECG trace and rheo-gram will correlate.

9). Existence of wave U on ECG trace is a criterion which demonstrates the most hazardous pathology. In most critical case, the amplitude of wave U is equal to amplitudes of waves P and T. Formation of wave U is a signal which manifests that the last protective function of cardiovascular system is actuated. From physical standpoint, myocardium is weakened, the life of the body is supported by pump function of aorta. As a rule, in this case wave S is not shown. This testifies to the fact that contractile function of myocardium is off.

10). To reveal the pathological symptoms and make them more obvious, the orthostatic test is made. It is a necessary procedure aimed at investigation of arterial pressure redistribution following the change of patient position from horizontal to upright. One to three minutes are required for pressure redistribution, two or three records should be made within this period. In normal condition of the body, the orthostatic test records do not differ, in case of pathological condition the amplitudes in pathological phases have some distinctions.

11). When diagnostics is completed, the patient receives the medical comment and recommendations on procedures to be made according to diagnosis.

9.2. Comparison analysis of diagnostics data: heart cycle phase analysis method data against classical multi-channel ECG registration data

The basic principles of ECG signal registration are as follows:

ECG standard leads

This method was offered by W. Einthoven in 1913 the leads fix a difference in the potentials between two points in electric field, far from the heart, in front plane on extremities.
A hypothetic line connecting two electrodes that participate in ECG lead production is called a lead axis. I, II, III.

Enhanced unipolar limb (Goldberger's) leads

This registration concept was offered by Goldberger in 1942, it involves the above method, but it is added by registration of a middle potential of two other extremities aV R , aV L a n d aV F.

Unipolar (Wilson's) chest leads

These leads were offered for registration by Wilson in 1934.
Following this idea, a difference in potentials is registered between the active positive electrodes placed at certain points on the chest surface and the unipolar negative Wilson's electrode.
Diagnostics criteria applicable to the classical multi-channel ECG registration are based on the following principles.

Sinus rhythm:

+ PII, it precedes QRS.
The same unchanged shape of P in one lead.

Atrial rhythm:

– PII,III – non-regular QRS complexes.

Rhythm from AV lead:

1. P is not available, P & QRS fusion (usually of unchanged shape).
2. P is registered after QRS complex.

Idioventricular rhythm:

1. A heart frequency < 40 beats/min.
2. A stretched and deformed QRS.
3. No connection between QRS and P is available.

Sinus tachycardia:

A heart frequency of 90–160 beats/min., up to 180 beats/min.

Sinus bradycardia:

A heart frequency of 59-40 beats/min.

Sinus arrhythmia:

Fluctuations of intervals R–R > 0.15 l in connection with breathing phases; there is an increase in heat frequency at inspiration.

Sick sinus node syndrome:

Persistent sinus bradycardia, non-sinus various rhythm patterns, block of sinus atrioventerticular node, tachycardia & bradycardia syndrome.

Slow "replacing" escape complexes/ rhythms:

A Heart frequency <= 60/min., non-sinus pacing lead, isolated non-sinus complexes; a complex preceding R–R is longer; another complex after ectopic complex is normal or shorter.

Accelerated ectopic rhythms:

A heart frequency of 90–130 beats/min.; atrial, AV, ventricular rhythms

Migration of supraventricular pacing lead:

1. A smooth change in the shape and IN the P wave polarity;
2. A change in PQ interval duration;
3. Non-sharply defined changes in RR intervals.

Paroxysmal tachycardia (PT):

A heart frequency of 140–250 beats/min.. A sudden beginning and a sudden end, the rhythm is regular.

Atrial PT:

Deformed, −, +P, normal shape of QRS (seldom with an aberration); an AV conductivity degradation may occur, including AV block, degree I–II with a total QRS loss.

PT AV:

A heart frequency of 140-220 beats/ min., − P in II, III, and in VF leads – a fusion after QRS complex; QRS complex is not stretched, seldom with an aberration.

Ventricular PT:

A heart frequency of 140–220 beats/min., the rhythm is often regular, QRS > 0,12", deformed, discordant segment RS–T. T–wave. Atrioventricular dissociation, the atrii excite in their rhythm.

Atrial fibrillation:

Atrial wave f with a frequency of 350–700 per minute, non-regular ventricular rhythm.
Atrial flutter fibrillation:
Atrial wave F with a frequency of 200–400 per minute, regular atrial rhythm, often with ventricular component – 2:1, 3:1, 5:1, etc..

ECG indications of hypertrophy of left ventricle:

1. An increase in an amplitude of R wave in left chest leads (V5, V6) and in an amplitude of S wave of right chest leads (V1, V2); in this case RV4<RV5 or RV4<RV6; RV5,6>25 mm or RV5,6+SV1>=35 mm (for persons aged over 40) and >=45 mm (for persons aged under 40).
2. Indications of rotation of the heart about the longitudinal axis counterclockwise: transition zone shifting to the right, into lead V2, loss of S waves in left chest leads (V5, V6);
3. Shifting of electric axis to the left. In this case R1>15mm; RAVL>=11 mm or R1 + S111 >= 25 mm.
4. Shifting of segment RS-T n leads V5, 6, 1, AVL below the isoline and formation of a negative or double-phase T-wave in leads 1, AVL and V5, V6;
5. An increase in duration of an interval of internal QRS deviation in left chest leads (V5, V6) by more than 0,05 sec..

ECG indications of hypertrophy of right ventricle:

1. Shifting of heart electrical axis to the right (angle α is over +1000)
2. An increase in the amplitude of R wave in right chest leads (RV1V2) by more than 7 mm and in the amplitude of S wave in left chest leads (V5, V6). In this case, the quantitative criteria may be the following: an amplitude of RV1+SV5,6 >= 10,5 mm;
3. QRS complex appeared in V1 of rSR or QR type;
4. Indications of rotation of the heart about its longitudinal axis clockwise: (transition zone shifting to the left, to leads V5, 6 and appearance of QRS-complex of RS-type in these leads);
5. Shifting of segment RS–T downwards and appearance of negative T waves in leads III, aVF, V1, 2;
6. An increase in duration of an interval of internal deviation in right chest leads (V1) by more than 0,03 sec.

A combined hypertrophy of both ventricles can be identified from individual indications available on ECG referring to the left and right ventricle, respectively.

ECG indications of coronary heart disease (CHD)

The most significant indications of CHD on the ECG are various changes in the shape and polarity of T wave.

A high T-wave in chest leads indicates either subendocardial ischemia of front wall or subepicardial, transmural or intramural ischemia of rear wall of the left ventricle (excepted for young persons: sometimes a high positive T is registered in chest leads).

A negative coronary T in chest leads is an indication of subepicardial, transmural or intramural ischemia of front wall of the left ventricle.

Two-phase +–, –+ T-waves are usually detected at borders of ischemic zone of intact myocardium.

The most significant indication of ischem-ic damage of myocar-dium is shifting of segment RS–T upwards or downwards, i.e., above or below isoline.

Shifting of segment RS–T upwards in chest leads. It indicates that subepicardial or transmural indications of damage of left ventricle front wall are available.

RS–T depression in chest leads. It indicates that either ischemic damage in sub-endocardial parts of front wall or trans-mural dam-age of rear wall of left ventricle is available.

The most significant indication of heart muscle necrosis is a pathological Q wave (in case of nontransmural necrosis), or it can be QS available (under transmural infarction).

Table 8.

ECG changes under AMI of different location.

MI localization	Leads	Changes detected on ECG
Anteroseptal myocardial infarction	V1-V3	Q or QS, +RS–T, –T
Apical myocardial infarction	V3, V4	Q or QS,+ RS–T, –T
Anterolateral myocardial infarction	I, AVL, V5, V6	Q, +RS–T, -T
Extensive myocardial infarction (anterior)	I, AVL, V1–V6. III, AVF	Q or QS, +RS–T, –T Reciprocal changes: –RS–T, +T (high)
Anterobasal myocardial infarction	V2\4-V2\6 V3\4-V3\6	Q or QS, +RS–T, –T
Postero-diaphragmatic myo-cardial infarction (inferior myocardial infarction)	III, AVF III, II AVF V1-V4	Q or QS, +RS–T, –T Reciprocal changes: –RS–T, +T (high)
Posterobasal myocardial infarction	V7-V9 V1-V3	Q or QS, +RS–T, –T
Posterolateral myocardial infarction	V5, V6, III, AVF V1-V3	Q, +RS–T, –T Reciprocal changes: an increase in R, –RS–T, +T (high)
Extensive posterior myo-cardial infarction	III, AVF, II, V5, V6, V7-9 V1-V3	Q or QS, +RS–T, –T Reciprocal changes: an increase in R, –RS–T, +T (high)

Development of acute myocardial infarction (AMI), staged development:

AMI acute stage is characterized by a rapid formation of pathological Q wave or QS complex that develops within 1–2 days, by shifting of segment RS–T upwards, above the isoline, by fusing, first, with a positive T wave, second, with a negative T-wave. Upon expiration of several days: RS–T comes close to the isoline.

Week 2 or 3: RS–T becomes isoelectric, the coronary wave T is sharply deepened and becomes symmetrical and sharp.

Subacute stage: the pathological wave Q or QS (necrosis) and the negative T are registered (ischemia); an amplitude of the said wave decreases step by step from the 20–25 day. RS–T is located in this case on the isoline unless an aneurysm is formed (a "frozen" RS–T increase).

Myocardial scarring stage: a pathological Q wave or QS is available for years or for the rest of lifetime, with a slightly negative flattened or positive T available.

Blood is supplied to the heart via three main coronary arteries as follows:

1. Ramus interventricularis anterior is responsible mainly for the proper blood supply to the anterior portion of the inter-ventricular septum, the apex and partly to the postero-diaphragmatic wall.

2. Ramus circumflexus is responsible for blood supply to the antero-apical, lateral and posterobasal portions of the left ventricle.

3. Arteria coronaria dextra is the blood supplier for the right ventricle, posterior portion of the interventricular septum, postero-diaphragmatic wall of the left ventricle and partly its posterobasal portions.

ECG indications of small-focus myocar-dial infarction, but of non-acute type

RS–T segment shifting upwards or downwards: above or below the isoline and (or) various pathological changes available in the T wave (frequently a negative coronary T).
These pathological changes can be found during 3–5 weeks from the beginning of myocardial infarction development, and sometimes even for a longer period of time.

ECG indications of myocardial ischemia under angina pectoris attack

Various changes in the T wave and (or) ischemic depression of RS–T below the iso-line that normalizes upon attack reduction.
Indications of vasospastic angina pectoris can be very good identified on ECG during 24-hour monitoring. Angina pectoris attacks and episodes of a transitor increase of RS–T (its depressions occur during rest, at night, irrespective of physical exercises). During the attack, there are no significant changes in heart frequency and artery pressure. Changes in RS–T at the beginning and the end of the attack occur rapidly, sharply. Indications of ischemia disappear either spontaneously or upon nitroglycerin intake irrespective of reflex increase in heart frequency. During the attacks heavy disturbances in the rhythm and conductivity may frequently occur.

ECG indications under aortic valve insufficiency (aortic decompensation)

Apparent indications of hypertrophy of left ventricle without its systolic overloading (RS–T changes).
–RS–T, flatness or –T occur only during the decompensation and during the development of cardiac insufficiency.

In case of "mitralization" of aortic decompensation, in addition to the indications of the left ventricle hypertrophy, there may occur also indications of hypertrophy of the right ventricle (P = mitral).

ECG shape under aortic mouth constriction (aortic stenosis).

Apparent indications of hypertrophy of the left ventricle with its systolic overloading (–RS–T and two-phase or –T in left chest leads). Indications of complete or incomplete left bundle-branch block (not always).
In order to make the comparative analysis more illustrative, all criteria are summarized in a Table below (Fig. 109).

Conclusions:

The above methods are not mutually exclusive; they are capable of adding each other. The phase analysis is a more advanced method because it makes possible to carry out differentiated diagnostics of all factors influencing the operation of the cardiovascular system as a whole. This method can be used to monitor the therapy process. Indirect measuring of hemody-namic volumetric parameters compares favorably with other conventional methods available.
Economic efficiency of application of this medical equipment should be mentioned, too, that is determined by the factors as follows: this analyzing instruments is easy in operation and is offered at a budget price. It makes it possible to obtain this unique instrument in the market.

Classical theory of heart diagnostics based on conventional multi-channel ECG.

Diagnostics object	Diagnostics criteria	ECG leads	Anatomical part to be diagnosticated
Heart rhythm.	1. Sinus-originated P-wave available; 2. RR variability; 3. PQ variability	I II III aVR aVL aVF V1 V2 V3 V4 V5 V6	III, aVF, V1, V2 → RA, RV
Electrocardiosignal conductivity.	Manifestations: 1. Block of bundle of His and bundle branches 2. Intraventricle conductivity disturbance 3. AB block		
Hypertrophy of atria and ventricles	P and QRS changes in different leads		I, V5, V6, aVL → LA, LV
Coronary heart disease. (Angina pectoris, myocardial infarction, postinfarction cardiosclerosis).	1. QRS changes. 2. Pathological Q wave. 3. ST segment change. 4. T wave change		

Method capabilities:
Differentiated qualitative evaluation.
Postfactum diagnostics. Forecasting hardly possible.

Prospects:
Metrological support cannot be provided, therefore, the potentialities of this method have been exhausted

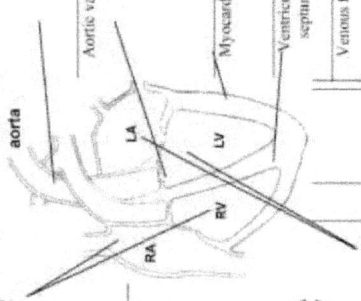

Theory of phase analysis of cardiovascular system based on single-channel ECG and point registration RHEO.

Anatomical part to be diagnosticated	Diagnostics object	Diagnostics criteria	Hemodynamic volume parameters to be identified
Aorta	Elasticity function. Aorta flabbiness available-not available.	Changes in arterial pressure amplitude on RHEO in phase J-T on ECG	SV - stroke volume of blood, ml
Aortic valves	Opening and closing functions	Changes in arterial pressure amplitude on RHEO in phase S-J on ECG	MV - minute stroke volume of blood, ml; PV_1 - volume of blood entering ventricle in diastole, ml; PV_2 - volume of blood entering ventricle in atrial systole, ml
Myocardium	Contraction function	R wave amplitude on ECG	PV_3 - volume of blood ejected by ventricle in rapid ejection phase, ml
Ventricular septum	Contraction function	S wave amplitude on ECG	PV_4 - volume of blood ejected by ventricle in slow ejection phase, ml
Venous flow	Fluidity function	Change in arterial pressure amplitude on RHEO in phase T-P0 on ECG	PV_5 - volume of blood (a share at SV) pumped by ascending aorta as peristaltic pump.

Capabilities of this method:
An integrated qualitative evaluation of anatomical part
+ Qualitative and quantitative evaluation of the heart functions
+ Measuring of hemodynamic parameters of the left heart and the aorta
+ Evaluation of hemodynamic parameters & identification of limit values and pathology, if any

Prospects:
Metrological support is available, and it is possible to determine & evaluate hemodynamics of the right heart, the actual blood flow structure in aorta and, in addition, 40 hemodynamic parameters including AP.

248

9.3. Comparative analysis of cardiovascular system diagnostics against background of lung diseases

This Chapter represents clinical data based on 21 cases of cardiovascular system diagnostics against the background of lung diseases. All these data are referred to in-hospital patients, and their respective Antecedent Anamneses prepared by classical standard methods are informative enough in order to perform a comparative analysis of two methods of diagnostics of cardiovascular system. Of concern to us are only diagnoses of cardiovascular conditions since the phase analysis of the heart cycle is not capable now of evaluating an effect of diseases of other organs on the respective phase characteristics reliably.

The intention of the developers is to compare data on duration of every phase of the respective heart cycle and both diagnosis options made. It should be noted that both methods used and the analyzer applied are certified. Therefore, the appropriate interpretation of differences in diagnosis options is of great importance.

Example No. 1 from Practice

Antecedent Anamnesis
Patient (full name: XX1), male, born 1965. Since 2005 registered were complaints about exudative pleurisy.

Diagnosis
Right-side exudative pleurisy of non-tuberculous etiology. Acute pericarditis

Clinical Examination
Complaints: dyspnea and weakness. Basic data: weight – 77 kgs, stature – 176 cm; body mass index (BMI) – 24.65kg/m2 (body mass norm).
Dynamic data: arterial pressure (AP) 100/60, 110/65, 100/75. Heart frequency: 90 beats per minute, breathing frequency: 19-22 breath movements per minute.
Auscultation: heart tones are damped down and rhythmic. Breathing.
Auscultation: right-side breathing in lower and rear segments is significantly weakened.
Percussion: the sound is dull on the right side in the lower segments of the lungs. Abdomen is soft, no pains available. The liver borders are extended.

Tests
Laboratory Testing.
General blood test: leukocytes – 4.8*10*9 (leukocyte formula shows no pathology), erythrocytes – 4.1*10*12, hemoglobin 140 g/l, erythrocyte sedimentation rate (ESR) – 3 mm/h, prothrombin – 106%, glucose – 6.9μmole/l, cholesterin – 5.1 μmole/l, C-reactive protein – negative, bi-lirubin – 7.8 μmole/l, AST 0,67, ALT 0.67.
Sputum analysis:

Bacterial analysis – multiple analysis (10 times), negative.
Secondary flora – gram-positive cocci
Urine analysis: no pathology.

Instrumentation testing.
ECG: P-0.8; PQ – 0.16; QRS – 0.10; QRST – 0.30; R–R – 0.55.
Heart frequency – 109 beats per minute.

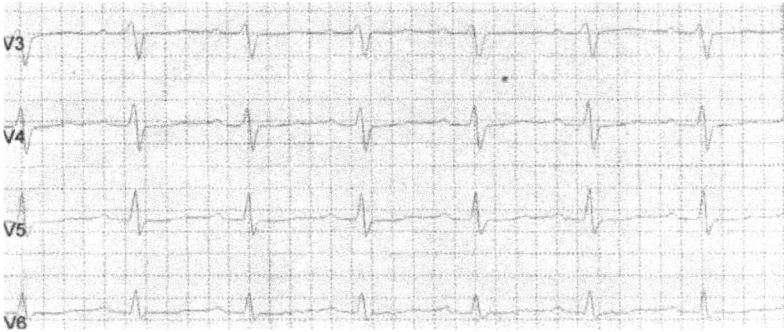

Apparent sinus tachycardia. Myocardial hypertrophy of the right ventricle (S-type). Repolarization disturbances in the antero-lateral wall of the left ventricle; the clinical picture may indicate coronary insufficiency (subendocardial insufficiency).

Radiographic examination:

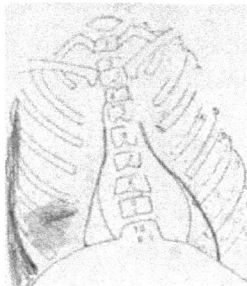

Right side: there is a section of an enhanced lung pattern in the middle lobe; paracostally and in sinus – pleural thickening. Left side: norm.

The heart is extended transversely.

Date of entry – 24.03.2005

Date of birth – 10.03.1965

Full name – XX1 24-03-2005 9-42-12

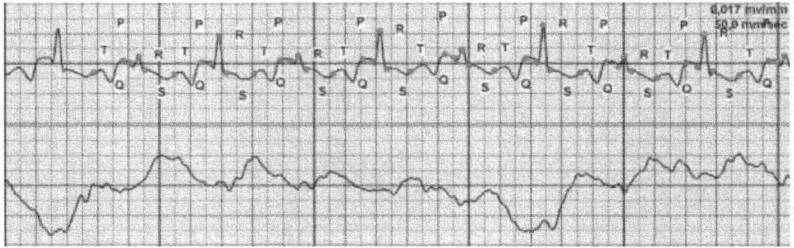

Diagnosis resulted from phase analysis.

The volume-related parameters of function of the heart and the associated blood vessels are within normal values.

Doctor's diagnosis.

The function of the aortic valve is weakened.

The ECG portion to the wave point S in the phase RS shows contraction of myocardium; the valve is leaking blood entering the aorta. This is the cause of an increase in the minimum arterial pressure.

Anatomic peculiarities of aortic valve in the tension phase indicated on the ECG curve. A changed condition indicated as smoothing of the wave SL is available. The wave S is not available. The aortic pressure in this phase is even slightly decreasing.

Anatomic peculiarities of the aortic valve in the full opening phase indicated on the ECG curve. A changed condition indicated as smoothing of the wave Lj is available. A minor increase in the aortic pressure is observed. A difference in pressures between the minimum and the maximum on the RHEO is small.

Flabbiness of aorta – not available.
Condition of elasticity of aorta – slightly reduced. The apex of the rheogram is flat.

Condition of contraction function of ventricular septum – norm.
Condition of contraction function of myocardium – the function is significantly diminished.

The amplitude of the lower portion of the QRS complex is a minimum, and that indicates the maximum passivity of the myocardium contraction.

Condition of venous blood flow – norm.

Condition of function of the lungs – the function of the lungs is significantly diminished.

A heavy modulation due to breathing on the rheogram curve is available.

Potential stroke problem – no sudden hemopulses are available.
Potential stroke problem is not available.

ORTHOSTATIC TEST WAS NOT PERFORMED.

Additional data. Irregular amplitudes of the QRS complex are detected. Every second complex has a very low amplitude. It indicates weakness of the heart, but hemodynamic parameters are within the norm.

SV (ml) – Stroke volume
| SV(ml)=62.55 %=0.00 | 39.32 | | 86.01 |

MV (l) – Minute volume
| MV(l)=7.13 %=0.00 | 4.48 | | 9.81 |

PV1 (ml) – Volume of blood in phase of rapid and slow filling of left (aortic) ventricle of heart
| PV1(ml)=18.26 %=0.00 | 14.50 | | 30.37 |

PV2 (ml) – Volume of blood flowing into left (aortic) ventricle of heart during systole
| PV2(ml)=44.29 %=0.00 | 24.82 | | 55.04 |

PV3 (ml) – Volume of blood ejected by left (aortic) ventricle of heart during rapid ejection phase
| PV3(ml)=37.15 %=0.00 | 23.33 | | 51.16 |

PV4 (ml) – Volume of blood ejected by left (aortic) ventricle of heart during slow ejection phase
| PV4(ml)=25.39 %=0.00 | 15.99 | | 34.85 |

PV5 (ml) – Volume of blood transferred by ascending
aorta functioning as peristaltic pump
| PV5(ml)=7.47 %=0.00 | 5.62 | | 8.45 |

The wave P on the ECG curve is practically not available. The RHEO pressure waves vary from cycle to cycle, and this is an indication of non-stability in blood flow. The RHEO pressure waves have a flattened apex indicating pathology in creation of the maximum arterial pressure.

	QRS	RS	QT	PQ	TT	SV (ml)	MV (1)	PV1 (ml)	PV2 (ml)	PV3 (ml)	PV4 (ml)	PV5 (ml)
Max	0.080	0.035	0.268	0.058	0.526	39.3	4.5	14.5	24.8	23.3	16.0	5.6
Min	0.110	0.060	0.268	0.058	0.526	86.0	9.8	31.0	55.0	51.2	34.9	8.4
Average	0.095	0.047	0.273	0.067	0.526	62.5	7.1	18.3	44.3	37.2	25.4	7.5
0	0.097	0.042	0.286	0.046	0.522	53.3	6.1	17.5	35.7	31.6	21.6	7.0
1	0.101	0.042	0.282	0.059	0.514	52.3	6.1	11.3	41.0	31.1	21.3	6.7
2	0.097	0.051	0.274	0.067	0.514	69.3	8.1	15.3	54.0	41.2	28.1	8.0
3	0.088	0.042	0.240	0.067	0.497	48.7	5.9	14.9	33.8	28.9	19.8	5.7
4	0.093	0.046	0.261	0.059	0.547	59.2	6.5	26.5	32.7	35.2	24.0	6.9
5	0.097	0.051	0.286	0.067	0.556	71.3	7.7	25.3	46.0	42.4	29.0	8.5
6	0.084	0.046	0.253	0.080	0.497	59.2	7.1	9.6	49.6	35.2	24.0	6.9
7	0.101	0.059	0.269	0.072	0.531	86.2	9.7	26.3	59.9	51.2	34.9	8.9
8	0.097	0.051	0.278	0.059	0.526	70.0	8.0	22.4	47.6	41.6	28.4	8.2
9	0.105	0.051	0.282	0.067	0.531	69.3	7.8	17.7	51.6	41.2	28.1	8.0
10	0.105	0.051	0.261	0.072	0.509	65.5	7.7	15.0	50.5	39.0	26.6	7.0
11	0.088	0.042	0.265	0.076	0.543	51.9	5.7	17.7	34.1	30.8	21.1	6.6
12	0.093	0.042	0.257	0.059	0.522	50.3	5.8	19.0	31.3	29.9	20.4	6.1
13	0.097	0.051	0.257	0.063	0.522	66.3	7.6	24.1	42.3	39.4	26.9	7.2
14	0.101	0.055	0.274	0.067	0.543	77.7	8.6	27.6	50.1	46.2	31.5	8.4
15	0.097	0.051	0.269	0.063	0.522	68.6	7.9	21.4	47.1	40.7	27.8	7.8
16	0.088	0.042	0.265	0.072	0.531	51.9	5.9	16.4	35.4	30.8	21.1	6.6
17	0.097	0.046	0.269	0.063	0.539	59.8	6.7	22.2	37.6	35.5	24.3	7.1
18	0.101	0.042	0.358	0.067	0.611	59.4	5.8	16.0	43.4	35.2	24.2	9.0
19	0.093	0.046	0.278	0.059	0.451	61.5	8.2	40.4	102.0	36.5	25.0	7.6
20	0.088	0.055	0.265	0.076	0.522	78.5	9.0	21.3	57.2	46.7	31.9	8.6
21	0.093	0.046	0.269	0.076	0.522	60.4	6.9	14.1	46.3	35.9	24.5	7.3
22	0.097	0.042	0.282	0.067	0.539	52.8	5.9	15.5	37.3	31.4	21.5	6.8
23	0.097	0.046	0.278	0.067	0.526	61.0	6.9	15.9	45.1	36.2	24.8	7.4
24	0.084	0.042	0.261	0.067	0.518	51.9	6.0	16.2	35.7	30.8	21.1	6.6
25	0.097	0.046	0.282	0.072	0.539	61.5	6.9	16.9	44.6	36.5	25.0	7.6
26	0.093	0.051	0.265	0.072	0.518	68.6	7.9	18.2	50.3	40.7	27.8	7.8
27	0.097	0.051	0.269	0.067	0.526	68.6	7.8	21.0	47.6	40.7	27.8	7.8

| 28 | 0.093 | 0.046 | 0.269 | 0.067 | 0.526 | 60.4 | 6.9 | 18.4 | 41.9 | 35.9 | 24.5 | 7.3 |
| 29 | 0.093 | 0.046 | 0.274 | 0.067 | 0.522 | 61.0 | 7.0 | 16.2 | 44.7 | 36.2 | 24.8 | 7.4 |

Comparative analysis of example No. 1 from practice

Phase duration.
A difference in the phase duration between the QRS complexes is 0.065 s.

A difference in the phase duration between the PQ phases is 0.102 s.

A difference in the phase duration between the QT phases is 0.027 s.

These differences are significant, and this may make difficult to ana-lyze both diagnoses by the doctor.

Diagnoses.
Classical diagnosis option is very short: myocardial hypertrophy of the right ventricle. Repolarization disturbances in the anterolateral wall of the left ventricle (possible is coronary insufficiency).
The phase analysis makes it possible to produce a more detailed analysis. The phase analysis is capable of identifying 10 functions of the cardiovascular system.
In order to understand better differences in making a diagnosis, let us carry out a close examination of the ECG curve shapes for both methods. The classical ECG multi-lead method is not capable of producing an informative and indicative ECG curve that can be produced by the analyzer CARDIOCODE. In order to make the diagnosis "hypertrophy of the right ventricle", used are indications of the multi-lead ECG curve mentioned in Chapter 9.2.

The phase analysis shows the following.

Differences in R-wave amplitude are available in one-cycle intervals, and that is an indication of a significant pathology rather in the central nervous system than in a local segment of the heart. Alternating sequences of differences in the R-wave amplitudes are of systematic nature. The wave S is not available. This indicates that there is a full loss in function of the contraction of the ventricle myocardium. The wave T is inverted. The wave P is not expressed. These signs show that there is a significant change in the shape of the heart caused by a factor that has not been identified yet. Therefore, it is required to analyze the relevant RHEO. First, the functioning of the aortic valve should be examined. The curves show that the arterial pressure in the phase of the valve tension does not increase. That is an indication of its sound condition. During the opening, in the phase of rapid blood ejection, the arterial pressure (AP) is increasing significantly. In the phase of slow blood ejection, there is a minor increase in the AP. This depends on the aorta elasticity. Therefore, the maximum values of the systolic pressures cannot be great, and that is supported by the relevant data at the date of the patient entry at the hospital, i.e., 100/60 mm Hg.

Therefore, the therapy target should be removal of the system problem connected with changes in the R wave amplitudes on the ECG curve, increasing at the same time elasticity of the aorta. Other problems should be treated as minor ones.

Example No. 2 from Practice

Antecedent Anamnesis
Patient (full name: XX2), male, born 1973. Since 2004 registered as tuberculous patient.
Abdomen is swollen, pains in the right hy-pochondrium. Liver borders are within the norm.
Urination is within the norm.

Diagnosis
Disseminated pulmonary tuberculosis, destruction & infiltration phase. IA, Tuberculosis Mycobacteria (TMB)+.

Clinical Examination
Complaints: dyspnea at a minor physical activity, weakness, cough.
Basic data: weight – 57 kgs, stature – 165 cm; body mass index (BMI) – 20.47kg/m2 (body mass norm).
Dynamic data: arterial pressure (AP) 10 0/6 0 , 110/6 0 .
Heart frequency: 92 beats per minute, breathing frequency: 19–21 breath movements per minute.
Auscultation: heart tones are damped down and rhythmic.
Breathing in lungs: weakened on both sides, random bubbling moist rales on both sides.

Tests
Laboratory Testing.
General blood test: leukocytes – 12.6*10*9, erythrocytes – 4.52*10*12, hemoglobin 95 g/l, erythrocyte sedimentation rate (ESR) – 50 mm/h, prothrombin – 106%, glucose –4.5 μmole/l; bilirubin, total – 5.1 μmole/l, bilirubin conjugated – 0.0; AST 0,33μmole/l.h, ALT 0.08 μmole/l.h. Bacterial inoculation: significant growth. Urine analysis: protein 0.033, Leucocytes-28 in visual field, bacteria.

Instrumentation testing.
ECG: P-0.10; PQ – 0.14; QRS – 0.08;
QRST – 0.36; R–R – 0.59.
Heart frequency – 101 beats per minute.

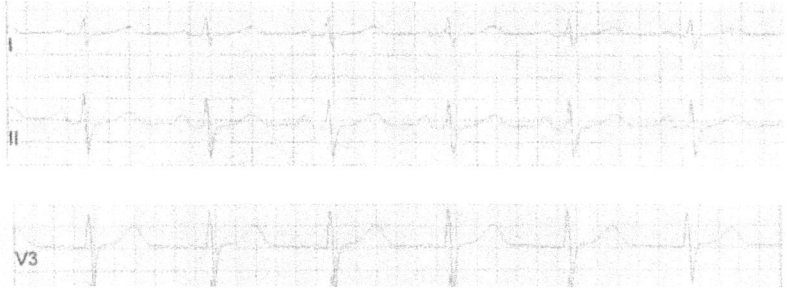

Apparent sinus tachycardia. Myocardial hypertrophy of the right atrium. Myocardial hypertrophy of the right ventricle (S-type). Diffuse changes in myocardium.

Radiographic examination:

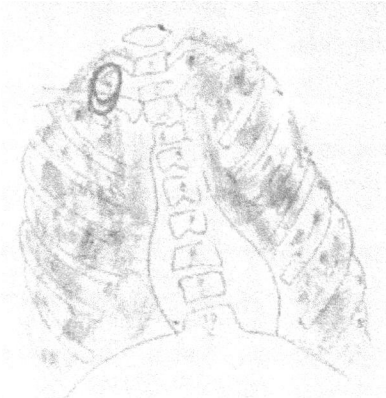

On both sides, throughout the length: scattered nidi without clear contours, multiple infiltrative foci; on the right side in S1: a destruction cavity – 3.5*2.0 cm.

Date of entry – 02.06.2005
Date of birth – 01.01.1973
Full name – XX2 02-06-05 09-25-46 am

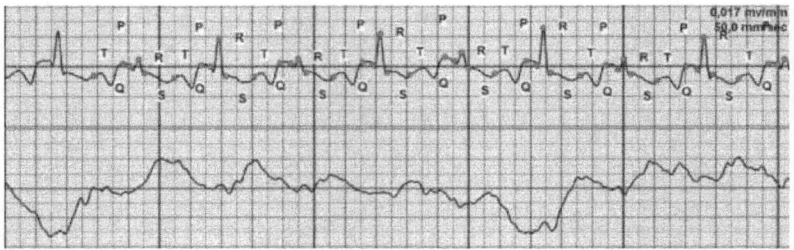

Diagnosis resulted from phase analysis.
The volume-related parameters of function of the heart and the associated blood vessels are within normal values.

Doctor's diagnosis.
Function of aortic valve – norm.

Anatomic peculiarities of aortic valve in tension phase indicated on the ECG curve – norm.

Anatomic peculiarities of aortic valve in the full opening phase indicated on the ECG curve. A changed condition indicated as smoothing of the wave Lj is available.

Flabbiness of aorta – not available.

Condition of elasticity of aorta. A weak pumping function of the aorta is available (s. phase analysis). The aortic pressure according to the RHEO cannot reach normal values. A difference in pressures between the maximum and the minimum is small. The apex of the rheogram is pyramidal and is left referred to the wave point T.

Condition of contraction function of ventricular septum – the function is diminished. The R wave point amplitude on the ECG is a half of norm.

Condition of contraction function of myocardium – the function is diminished. A bifurcation of the S wave is periodically detected.

Condition of venous blood flow – norm.

Condition of function of the lungs – the function of the lungs is diminished.

A minor modulation due to breathing both on the rheogram and the ECG is available.

Potential stroke problem – no sudden hemopulses are available.

Potential stroke problem is not available.

ORTHOSTATIC TEST WAS NOT PERFORMED.

SV (ml) – Stroke volume

| SV(ml)=63.96 %=0.00 | 39.55 | | 86.81 |

MV (l) – Minute volume

| MV(l)=7.13 %=0.00 | 4.41 | | 9.67 |

PV1 (ml) – Volume of blood in phase of rapid and slow filling of left (aortic) ventricle of heart

| PV1(ml)=25.44 %=0.00 | 15.60 | | 33.48 |

PV2 (ml) – Volume of blood flowing into left (aortic) ventricle of heart during systole

| PV2(ml)=38.52 %=0.00 | 23.95 | | 53.34 |

PV3 (ml) – Volume of blood ejected by left (aortic) ventricle of heart during rapid ejection phase

| PV3(ml)=37.97 %=0.00 | 23.47 | | 51.63 |

PV4 (ml) – Volume of blood ejected by left (aortic) ventricle of heart during slow ejection phase

| PV4(ml)=25.93 %=0.00 | 16.08 | | 35.18 |

PV5 (ml) – Volume of blood transferred by ascending
aorta functioning as peristaltic pump

| PV5(ml)=8.57 %=0.00 | 5.70 | | 8.62 |

	QRS	RS	QT	PQ	TT	SV (ml)	MV (1)	PVl (ml)	PV2 (ml)	PV3 (ml)	PV4 (ml)	PV5 (ml)
Min	0.080	0.035	0.272	0.059	0.539	39.5	4.4	15.6	23.9	23.5	16.1	5.7
Max	0.110	0.060	0.272	0.059	0.539	86.8	9.7	33.5	53.3	51.6	35.2	8.6
Average	0.091	0.046	0.307	0.036	0.539	64.0	7.1	25.4	38.5	38.0	26.0	8.6
0	0.089	0.046	0.304	0.038	0.519	65.4	7.6	20.0	45.4	38.8	26.6	8.7
1	0.093	0.046	0.312	0.034	0.531	65.9	7.4	23.4	42.5	39.1	26.8	8.8
2	0.097	0.051	0.308	0.034	0.531	74.7	8.4	28.1	46.6	44.3	30.3	9.4
3	0.101	0.055	0.308	0.038	0.536	84.2	9.4	30.7	53.4	50.0	34.2	10.1
4	0.089	0.042	0.304	0.038	0.531	56.0	6.3	20.4	35.5	33.2	22.8	7.8
5	0.089	0.042	0.304	0.038	0.531	56.0	6.3	20.4	35.5	33.2	22.8	7.8
6	0.089	0.042	0.304	0.038	0.531	56.0	6.3	20.4	35.5	33.2	22.8	7.8
7	0.093	0.046	0.304	0.034	0.527	64.9	7.4	24.4	40.5	38.5	26.4	8.6
8	0.093	0.046	0.308	0.034	0.540	65.4	7.3	27.0	38.4	38.8	26.6	8.7
9	0.089	0.042	0.312	0.034	0.552	56.8	6.2	25.3	31.5	33.7	23.1	8.1
10	0.084	0.046	0.308	0.038	0.561	66.3	7.1	32.1	34.3	39.4	27.0	9.0
11	0.093	0.046	0.308	0.038	0.557	65.4	7.0	29.3	36.1	38.8	26.6	8.7
12	0.089	0.042	0.304	0.038	0.552	56.0	6.1	25.1	30.9	33.2	22.8	7.8

Comparative analysis of example No. 2 from practice

Phase duration.

A difference in the phase duration between the QRS complexes is 0.011 s.

A difference in the phase duration between the PQ phases is 0.02 s. (insignificant).

A difference in the phase duration between the QT phases is 0.053 s. (insignificant).

A significant difference is available only in the duration between the QRS complexes.

Diagnoses.

Classical diagnosis option: hypertrophy of myocardium of the right segments of the heart.

The phase analysis shows the following.

Visually, a modulation of the ECG and the RHEO by a breathing wave is detected. This is indication No.1 of a significant weakness of the lungs. The arterial pressure in the slow ejection phase is actually not increasing. This is supported by pressure gauge measuring: the corresponding reading is 100/60 mm Hg. It should be noted that there is a great amplitude of the T wave available. It is an indication of aorta loading. This fact can be supported by the relevant value PV5.
The salient feature of the diagnosis is that there is no increase in the arterial pressure available despite the fact that the aorta is loaded

in the slow ejection phase. A bifurcation of the S wave shows that hypertrophy of the ventricles is available. All above indications support the fact that the function of the lungs is weak, and this leads to a deficit in oxygen supplied to blood cells.

Therefore, the therapy target should be the proper treatment of the lungs in order to remove their pathology.

Example No. 3 from Practice

Antecedent Anamnesis
Patient (full name: XX3), male, born 1962. Since 1999 registered as pulmonary tuberculous patient.

Diagnosis
Pulmonary fibrous cavernous tuberculosis, infiltration phase, IIB, Tuberculosis Mycobacteria (TMB)+.

Clinical Examination
Complaints: dyspnea at physical activity, weakness and cough accompanied by white sputum, hyperhidrosis. Basic data: weight – 57 kgs, stature – 165 cm; body mass index (BMI) – 18.31 kg/m2 (body mass deficit). Dynamic data: arterial pressure (AP) 110/70, 12 0/8 0. Heart frequency: 80 beats per minute, breathing frequency: 19–20 breath movements per minute.

Auscultation: heart tones are damped down and rhythmic. Breathing in lungs: rough on both sides, single dry rales on both sides. Abdomen: no pains available. Liver borders: the liver border is at the edge of the costal margin. Urination is within the norm.

Tests
Laboratory Testing.

General blood test: leukocytes – 6.7*10*9, erythrocytes – 4.42*10*12, hemoglobin 144 g/l, erythrocyte sedimentation rate (ESR) – 28 mm/h, prothrombin – 89%, glucose –5.0 μmole/l; bilirubin, total – 5.8 μmole/l, bilirubin conjugated – 0.0; AST μmole/l.h – 0.50, ALT 0.67 μmole/l.h. Bacterial inoculation – growth, 4 colonies. Urine analysis: no pathology available.

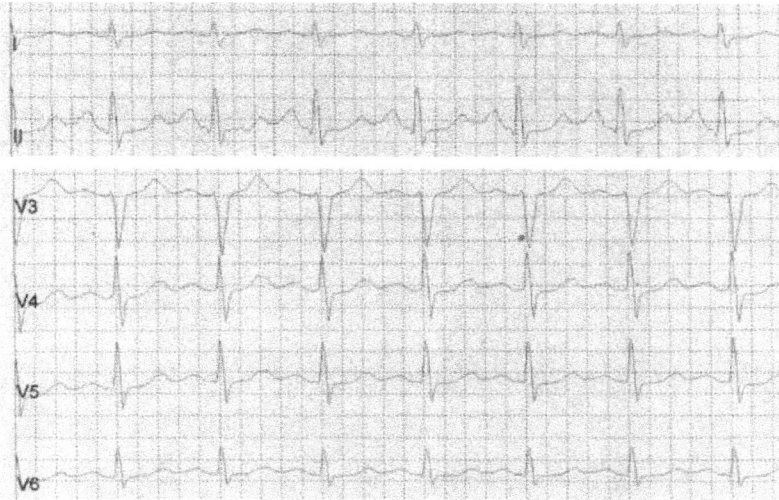

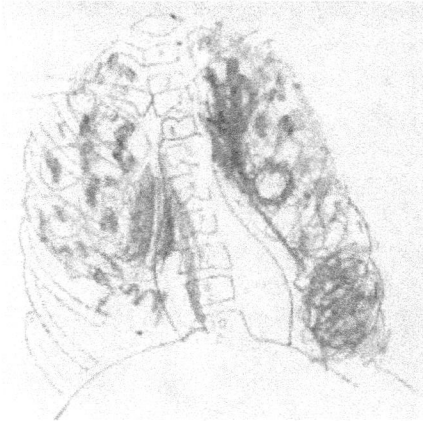

Instrumentation testing.

ECG: P-0.10; PQ – 0.13; QRS – 0.10; QRST – 0.34; R–R – 0.51; Heart frequency – 118 beats per minute.

Apparent sinus tachycardia. Myocardial hypertrophy of the right atrium. Myocar-dial hypertrophy of the right ventricle (S-type). Repolarization disturbances in the anterolateral wall of the left ventricle may be interpreted as an indication of subendo-cardial insufficiency.

Radiographic examination:

Right side, up to rib 5 and throughout the length on the left side: apparent ring fi-brosis, multiple polymorphous nidi; left in S6: a cavity 2,0 cm with infiltrated walls; in the bottom lobe of lungs: an in-homogenous focus without clear contours of 5.0*3.0 cm, with destruction areas; the roots of lungs are expanded, thickened, the left root is directed upwards. The heart is shifted to the left.
Radiographic examination:

Right side, up to rib 5 and throughout the length on the left side: apparent ring fi-brosis, multiple polymorphous nidi; left in S6: a cavity 2,0 cm with infiltrated walls; in the bottom lobe of lungs: an in-homogenous focus without clear contours of 5.0*3.0 cm, with destruction areas; the roots of lungs are expanded, thickened, the left root is directed upwards. The heart is shifted to the left.

Date of entry – 14.06.2005
Date of birth – 13.10.1962
Full name – XX3 14-06-05 10-48-13 am

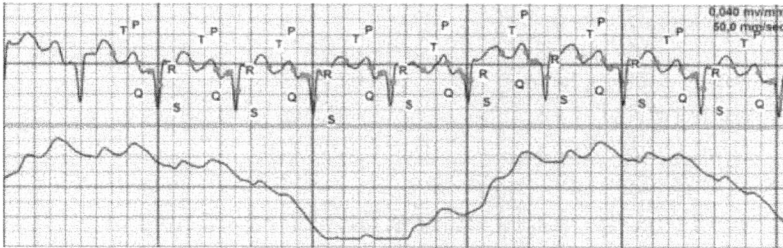

Diagnosis resulted from phase analysis.
SV (ml.l) – Within the norm
MV (l) – Within the norm
PV2 (l) – An increased loading on the left atrium by blood volume in the atrium systole is available.
PV3 (l) – Within the norm
PV4 (l) – Within the norm
PV5 (l) – The volume of blood pumped by ascending aorta as peristaltic pump is increased.
The tonus of ascending aorta is strengthened. A consequence is a reduced loading on the
left ventricle due to a decreased resistance of the aorta.
The volume-related parameters of function of the heart and the associated blood vessels
are within normal values.

Doctor's diagnosis.
Function of aortic valve – norm.

Anatomic peculiarities of the aortic valve in the tension phase indicated on the ECG curve. A changed condition as an increase in the amplitude of the wave SL that exceeds the wave R.

Anatomic peculiarities of aortic valve in the full opening phase indicated on the ECG curve. There is no pressure increase available in this phase at all. This is indicative of significant pathology.

Flabbiness of aorta – apparently available.

There is no increase, but a sharp decrease in the pressure in the slow ejection phase available.

Condition of elasticity of aorta – norm.

The apex of the rheogram is parabolic and is left of the wave point T. An increase in the aortic pressure is provided only due to its own pumping function.

Condition of contraction function of ventricular septum – the function is considerably diminished. The R wave amplitude is less than that of the T wave.

Condition of venous blood flow – the venous blood flow is heavily hindered.

There is an increase in the pressure on the RHEO after dicrotic valley available, and the said increase exceeds the maximum pressure which is available in the systole.

Condition of function of the lungs – the function of the lungs is considerably diminished. A significant modulation by breathing on the rheogram curve is available.

Potential stroke problem – no sudden hemopulses are available. Potential stroke problem is not available.

ORTHOSTATIC TEST WAS NOT PERFORMED.

Additional information. Blood flow is provided mainly by the pumping function of the aorta. But it exceeds the norm (s. phase analysis). There is also a loading on the ventricles available due to weakness of the ventricular septum.

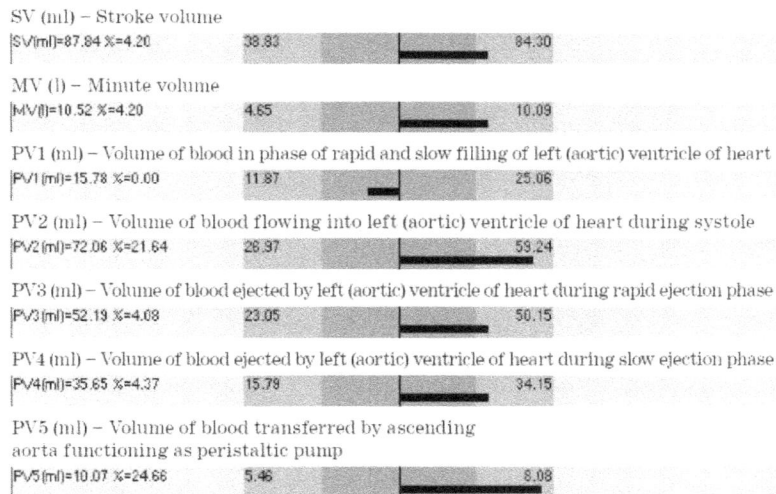

SV (ml) – Stroke volume
SV(ml)=87.84 %=4.20   38.83   84.30

MV (l) – Minute volume
MV(l)=10.52 %=4.20   4.65   10.09

PV1 (ml) – Volume of blood in phase of rapid and slow filling of left (aortic) ventricle of heart
PV1(ml)=15.78 %=0.00   11.87   25.06

PV2 (ml) – Volume of blood flowing into left (aortic) ventricle of heart during systole
PV2(ml)=72.06 %=21.64   28.37   59.24

PV3 (ml) – Volume of blood ejected by left (aortic) ventricle of heart during rapid ejection phase
PV3(ml)=52.19 %=4.08   23.05   50.15

PV4 (ml) – Volume of blood ejected by left (aortic) ventricle of heart during slow ejection phase
PV4(ml)=35.65 %=4.37   15.78   34.15

PV5 (ml) – Volume of blood transferred by ascending aorta functioning as peristaltic pump
PV5(ml)=10.07 %=24.66   5.46   8.08

	QRS	RS	QT	PQ	TT	SV (ml)	MV (1)	PV1 (ml)	PV2 (ml)	PV3 (ml)	PV4 (ml)	PV5 (ml)
Min	0.080	0.035	0.262	0.055	0.501	38.8	4.6	11.9	27.0	23.0	15.8	5.5
Max	0.110	0.060	0.262	0.055	0.501	84.3	10.1	25.1	59.2	50.1	34.2	8.1
Ave rage	0.081	0.057	0.280	0.060	0.501	87.8	10.5	15.8	72.1	52.2	35.6	10.1
0	0.072	0.055	0.269	0.072	0.505	82.4	9.8	16.2	66.2	49.0	33.4	9.6
1	0.072	0.051	0.274	0.067	0.505	73.2	8.7	14.4	58.8	43.5	29.7	9.0
2	0.080	0.046	0.278	0.059	0.497	63.1	7.6	9.9	53.2	37.5	25.6	8.1
3	0.084	0.063	0.282	0.059	0.505	103.1	12.2	21.8	81.3	61.3	41.8	11.2
4	0.072	0.046	0.274	0.076	0.505	63.7	7.6	7.1	56.6	37.8	25.9	8.2
5	0.114	0.080	0.320	0.025	0.505	151.1	17.9	33.0	118.1	89.9	61.2	14.6
6	0.101	0.067	0.299	0.042	0.497	113.9	13.8	15.8	98.1	67.7	46.2	11.9
7	0.067	0.051	0.261	0.076	0.501	71.9	8.6	13.7	58.2	42.7	29.2	8.7
8	0.072	0.051	0.265	0.072	0.501	71.9	8.6	13.7	58.2	42.7	29.2	8.7
9	0.080	0.051	0.269	0.063	0.497	71.3	8.6	13.6	57.6	42.3	28.9	8.5
10	0.088	0.055	0.290	0.055	0.514	83.1	9.7	19.1	64.0	49.4	33.7	9.8
11	0.093	0.076	0.295	0.051	0.497	137.8	16.6	13.6	124.2	82.0	55.9	13.6
12	0.076	0.051	0.274	0.067	0.497	72.6	8.8	8.6	63.9	43.1	29.5	8.9
13	0.067	0.046	0.265	0.072	0.497	63.1	7.6	9.9	53.2	37.5	25.6	8.1
14	0.084	0.067	0.278	0.051	0.497	112.8	13.6	30.8	82.0	67.1	45.7	11.6

Comparative analysis of example No. 3 from practice

Phase duration.

A difference in the phase duration between the QRS complexes is 0.019 s.
A difference in the phase duration between the PQ phases is 0.07 s. (significant).
A difference in the phase duration between the QT phases is 0.06 s. (significant).

Diagnoses.

Classical diagnosis option: hypertrophy of myocardium of the right segments of the heart.

The phase analysis shows the following:

At first, it should be noted that there is a very small R wave amplitude is available. This is an indication of a heavy problem of function of the ventricular septum. As a rule, it is manifested in weakness and vertigo, therefore a patient tends to take a horizontal position. At the same time, the most notice is attracted by the great amplitude of the wave T. It is indicative of loading available on the pumping function of the aorta. The wave P is hypertrophied. It is an indication of the atrial hypertrophy. Then, when studying the RHEO, the curve shows a well-defined dip instead of an apex in the slow ejection phase, while the arterial pressure must increase in this phase. This is an indication of a considerable flabbiness of aorta.

The volume of blood entering the aorta in the systole does not produce the required pressure since the aortic wall is extremely stretched-out and does not show the proper elasticity. This is the main cause of a decrease in the arterial pressure detected. This fact is manifested in vertigo. Besides, this occurrence is aggravated by

273

the small R wave amplitude that makes more difficult diagnostics and therapy. So, the pathology cause is only swelling of the membrane cells of the ventricle septum and the aorta. In addition, it should be mentioned that the RHEO is modulated by a breathing wave. This is an indication of lung weakness.

The data generated by automatic analysis support us in diagnostics very well. So, it is evident that all parameters exceed the relevant norms, with the exception of the parameter PV1. Of particular interest could be the parameter PV 2 characterizing the contraction function of the ventricles and the parameter PV5 responsible for pumping function of the aorta. Both parameters in question show that the ventricle is overloaded and, as a consequence, the aorta requires much more power to maintain the blood pumping.

The target of therapy should be to maintain the lung function and remove swelling of the membrane cells of the heart and the associated blood vessels.

Example No. 4 from Practice

Antecedent Anamnesis
Patient (full name: XX4), male, born 1974. Since 1998 registered as pulmonary tuberculous patient.
Diagnosis

Pulmonary fibrous cavernous tuberculosis, infiltration & seeding phase, IIB, TMB+, chronic cardiopulmonary decompensation, degree II, III. Renal amyloidosis.
Clinical Examination

Complaints: dyspnea at a minor physical activity, weakness, cough accompanied by grey sputum, periodically edema of feet, legs & hands, arms and face. Basic data: weight – 57 kgs, stature – 165 cm; body mass index (BMI) – 16.31 kg/m2 (body mass deficit).
Face: pasteous; edema of feet, legs & hands.
Dynamic data: arterial pressure (AP) 110/65, 110/70, 120/80. Heart frequency: 80 beats per minute, breathing frequency: 21-22 breath movements per minute.
Auscultation: heart tones are damped down and rhythmic.
Breathing in lungs: rough on both sides, dry & moist rales on both sides. Abdomen: soft, no pains available. Liver borders: the liver border extends 3 cm beyond the edge of the costal margin. Urination: abnormal frequency.

Tests
Laboratory Testing.
General blood test: leukocytes – 9.4*10*9, erythrocytes – 4.52*10*12, hemoglobin 102 g/l, erythrocyte sedimentation rate (ESR) – 76 mm/h, prothrombin – 106%, urea – 25.8 μmole/l; glucose –5.0; bilirubin, total – 4.3, bilirubin conjugated – 0.0; AST – 0,33 μmole/l.h, ALT 0.67 μmole/l.h; bacterial inoculation – rich

growth of Tuberculosis Mycobacteria (TMB). Urine analysis: protein 3.3, fresh erythro-cytes, granular, cylinder. Erythrocytes – 4–5 in visual field.

Instrumentation testing. ECG: P-0.08; PQ – 0.14; QRS – 0.08; QRST – 0.38; R–R – 0.78; heart frequency – 77 beats per minute. Sinus rhythm is available. Diffuse changes in myocardium are detected.

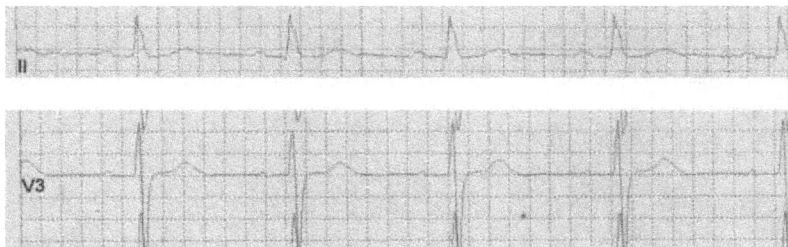

Radiographic examination:

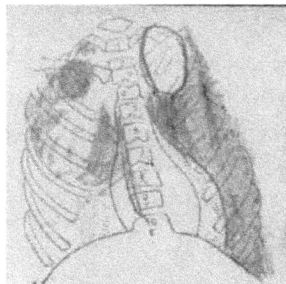

Right side, up to rib 5: subcortically, polymorphous nidi are available; in S2: a focus of 2.5*2.0 is inhomogeneous, the root is expanded. The left lung volume is reduced, homogeneously shadowed; in S1-2: a cavity of 6.0*4.0 cm; mediastina are shifted to the left.

Date of entry – 03.06.2005
Date of birth – 10.10.1974
Full name – XX4 03-06-05 09-32-06 am

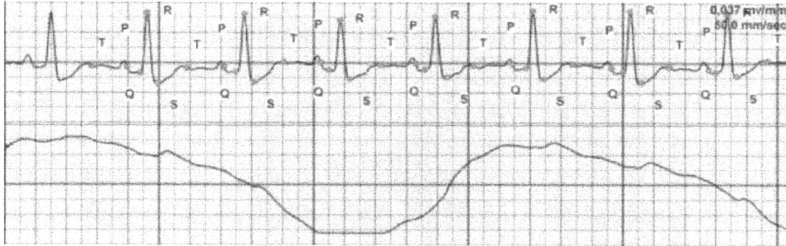

Diagnosis resulted from phase analysis.
Automatically generated diagnosis.

SV (ml.l) – An increased volume of blood enters the left heart. The heart is loaded by the blood volume. Possible is an increase in blood pressure in lesser circulation. The parameter PV5 should be thoroughly evaluated which is an indication of the blood volume pumped by the aorta as peristaltic pump.

MV (l) - An increased volume of blood enters the left heart. The heart is loaded by the blood volume. Possible is an increase in blood pressure in lesser circulation. The parameter PV5 should be thoroughly evaluated which is an indication of the blood volume pumped by the aorta as peristaltic pump.

PV1 (l) - An increased volume of blood enters the left ventricle. Possible is an increase in blood pressure in lesser circulation. This parameter measured depends on the functional state of the lungs. Therefore, the function of the lungs should be closely studied with use of additional instrumentation beyond this scope of supply of this analyzer. PV2 (l) – There is an increase in loading by the blood volume on the left atrium in the atrial systole. Possible is an increase in blood pressure in lesser circulation. A heavy loading by the volume is available (hyperfunction).

277

PV3 (l) – There is an increase in the blood volume ejected by the left ventricle in the rapid ejection phase available. The arterial pressure in greater circulation should be closely studied. Possible is an increase in the arterial pressure.

PV4 (l) – There is an increase in the blood volume ejected by the left ventricle in the slow ejection phase. The arterial pressure in greater circulation should be closely studied. Possible is an increase in the arterial pressure.

PV5 (l) – There is an increase in the blood volume pumped by the ascending aorta as peristaltic pump. The tonus of the ascending aorta is strengthened. The consequence is a decrease in loading on the left ventricle due to a reduced resistance by the aorta.

Doctor's diagnosis.

Function of aortic valve – norm.

Anatomic peculiarities of the aortic valve in the tension phase indicated on the ECG curve. A changed condition as a smoothed wave SL is available. The wave S is not available at all, and this wave is registered as a straight line.

Anatomic peculiarities of aortic valve in the full opening phase indicated on the ECG curve. A changed condition as a smoothed wave Lj is available. There is no pressure in the aorta produced in this phase at all. Flabbiness of aorta – not available. Condition of elasticity of aorta – norm.

The apex of the rheogram is parabolic and is left of the wave point T. Condition of contraction function of ventricular septum – norm. Condition of contraction function of myocardium – considerably diminished. The amplitude of the lower portion of the QRS complex is a minimum, and that is an indication of the maximum passivity of the myocardium contraction. Considering at the same time the wave S, it should be interpreted as a significant pathology of the myocardium and the aortic valve (s. phase analysis). Condition of venous blood flow – norm.

Condition of function of the lungs – the function of the lungs is considerably diminished. A significant modulation of the rheogram curve by breathing is available. Potential stroke problem – no sudden hemopulses are available. Potential stroke problem is not available.

ORTHOSTATIC TEST WAS NOT PERFORMED.

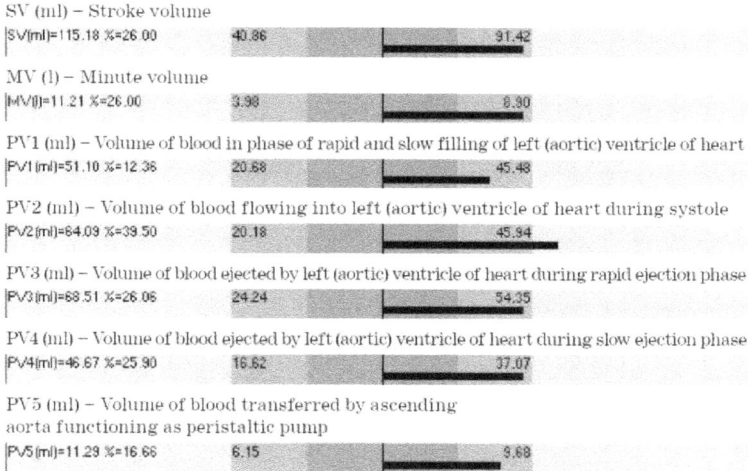

SV (ml) – Stroke volume

| SV(ml)=115.18 %=26.00 | 40.86 | 91.42 |

MV (l) – Minute volume

| MV(l)=11.21 %=26.00 | 3.98 | 8.30 |

PV1 (ml) – Volume of blood in phase of rapid and slow filling of left (aortic) ventricle of heart

| PV1 (ml)=51.10 %=12.36 | 20.68 | 45.48 |

PV2 (ml) – Volume of blood flowing into left (aortic) ventricle of heart during systole

| PV2 (ml)=64.09 %=39.50 | 20.18 | 45.94 |

PV3 (ml) – Volume of blood ejected by left (aortic) ventricle of heart during rapid ejection phase

| PV3 (ml)=68.51 %=26.06 | 24.24 | 54.35 |

PV4 (ml) – Volume of blood ejected by left (aortic) ventricle of heart during slow ejection phase

| PV4(ml)=46.67 %=25.90 | 16.62 | 37.07 |

PV5 (ml) – Volume of blood transferred by ascending
aorta functioning as peristaltic pump

| PV5(ml)=11.29 %=16.66 | 6.15 | 9.68 |

	QRS	RS	QT	PQ	TT	SV (ml)	MV (1)	PV1 (ml)	PV2 (ml)	PV3 (ml)	PV4 (ml)	PV5 (ml)	HR
Min	0.080	0.035	0.291	0.066	0.617	40.9	4.0	20.7	20.2	24.2	16.6	6.2	97.3
Max	0.110	0.060	0.291	0.066	0.617	91.4	8.9	45.5	45.9	54.3	37.1	9.7	97.3
Average	0.128	0.069	0.311	0.066	0.617	115.2	11.2	51.1	64.1	68.5	46.7	11.3	97.3
0	0.109	0.063	0.299	0.084	0.615	101.2	9.9	41.9	59.3	60.2	41.0	10.7	97.6
1	0.122	0.063	0.312	0.072	0.628	101.2	9.7	44.9	56.3	60.2	41.0	10.7	95.6
2	0.139	0.067	0.333	0.055	0.628	112.9	10.8	50.1	62.8	67.1	45.8	11.6	95.6
3	0,147	0,080	0,312	0,059	0,598	134,8	13,5	55,7	79,1	80,3	54,5	11,4	100,3
4	0,114	0,067	0,299	0,084	0,644	110,6	10,3	52,5	58,1	65,8	44,8	11,1	93,1
5	0,122	0,076	0,299	0,076	0,628	129,2	12,4	61,1	68,1	76,9	52,3	11,8	95,8
6	0,135	0,063	0,333	0,046	0,636	103,2	9,7	50,9	52,3	61,3	41,9	11,2	94,4
7	0,143	0,076	0,328	0,055	0,615	132,3	12,9	57,0	75,3	78,7	53,6	12,4	97,6
8	0,114	0,067	0,295	0,076	0,598	109,4	11,0	45,6	63,8	65,1	44,3	10,9	100,3
9	0,147	0,067	0,337	0,046	0,619	111,8	10,8	48,8	63,0	66,5	45,3	11,4	96,9
10	0,122	0,072	0,312	0,067	0,623	122,6	11,8	56,9	65,7	72,9	49,7	12,1	96,3
11	0,122	0,072	0,274	0,072	0,569	109,7	11,6	44,6	65,1	65,3	44,4	9,4	105,5

Comparative analysis of example No. 4 from practice

Phase duration.

A difference in the phase duration between the QRS complexes is 0.048s.
A difference in the phase duration between the PQ phases is 0.074s.
A difference in the phase duration between the QT phases is 0.08s.

Diagnoses.

Classical diagnosis option: diffuse changes in myocardium.
The phase analysis shows the following.
At first, it should be noted that there are a great scope of the diagnoses description produced automatically. The parameters analyzed show that the patient must be directed to a hospital. The most critical parameter indicating a dangerous health condition is the parameter exceeding 30% of the normal parameter value. In this case, we have to do with PV2 = 39,5 of the normal value. Other parameters also show deviations close to 30% of the value.
Special attention should be given to another two factors, when analyzing the ECG and RHEO: at first, the wave S is not available at all. After this wave, there is no phase development available from the point S to the Point TH. The wave T does not show a well-defined dip. Actually, there is no contraction of the ventricles is available, and the valve starts its opening practically from the wave point S. The RHEO curve supports this fact. The second factor in question is a heavy modulation of the RHEO by breathing. This is an indication of an extremely weak function of the lungs. The curve also shows that there is a rapid decrease in the AP available in the slow ejection phase.
These diagnoses describe an aggravated pathological status of the patient. The therapy recommendations could be only an appropri-

ate in-hospital therapy and treatment procedures capable of improving the general health status of the patient. The fact definitely supporting the accuracy of the above diagnoses is that the above mentioned patient died 5 months later after the described diagnostic procedure. Cause of death: cardiopulmonary decompensation.

Example No. 5 from Practice

Antecedent Anamnesis
Patient (full name: XX5), male, born 1958. Since 2001 registered as pulmonary tuberculous patient.

Diagnosis
Pulmonary fibrous cavernous tuberculosis of the left lung, infiltration phase, IIA, TMB+. Essential hypertension, degree I. Duodenal ulcer.

Clinical Examination
Complaints: hemoptysis, dyspnea at physical activity, cough accompanied by yellow sputum, hyperhidrosis. Basic data: weight – 57 kgs, stature – 165 cm; body mass index (BMI) – 18.31kg/m2 (body mass deficit).
Dynamic data: arterial pressure (AP) 13 0/8 0, 14 0/ 9 0.
Heart frequency: 84 beats per minute, breathing frequency: 22–24 breath movements per minute.
Auscultation: heart tones are damped down and rhythmic.
Breathing in lungs: rough on both sides, dry rales on the left side.
Abdomen: soft, no pains available. Liver borders: the liver border is at the edge of the costal margin.

Tests
Laboratory Testing. General blood test: leukocytes – $8.0*10*9$, erythrocytes – $4.3*10*12$, hemoglobin 115 g/l, erythrocyte sedimentation rate (ESR) – 30 mm/h, prothrombin – 106%,
glucose – 5.3 μmole/l; bilirubin, total – 6.8, bilirubin conjugated – 0.0; AST – 0,17 μmole/l.h, ALT – 0.17 μmole/l.h. Bacterial inoculation of sputum – rich TMB growth. Urine analysis: no pathology available. Instrumentation testing. ECG: P-0.10; PQ – 0.14; QRS – 0.09;

QRST – 0.32; R–R – 0.49; Heart frequency – 123 beats per minute.

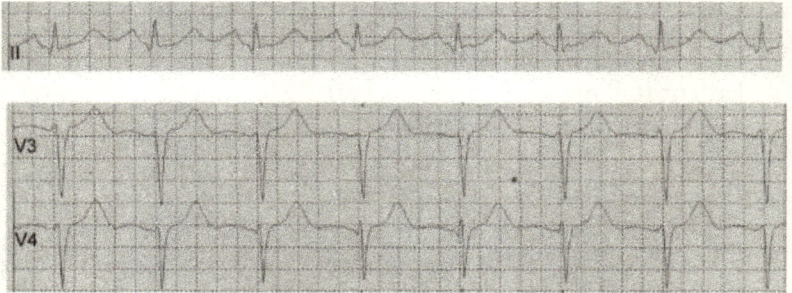

Apparent sinus tachycardia. Myocardial hypertrophy of the right atrium. Typical signs of scarry changes within the an-teroseptal area. Diffuse changes in myocardium; if clinical picture is available, possible as an expression of coronary insufficiency.

Radiographic examination:

Right side, SI and SII: nidi without clear contours are available, small cavities of destruction. Left side: the lung volume is reduced, throughout its length nidi without clear contours are available; there is a cavern of 4*2.5 cm available in Si-II; there is a focus of 3.5*2.5 cm without clear contours in S6 available. The heart is shifted to the left.

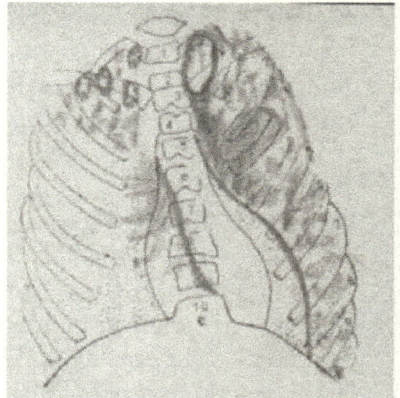

Date of entry – 03.06.2005
Date of birth – 01.01.1958
Full name – XX5 03-06-05 09-51-53 am

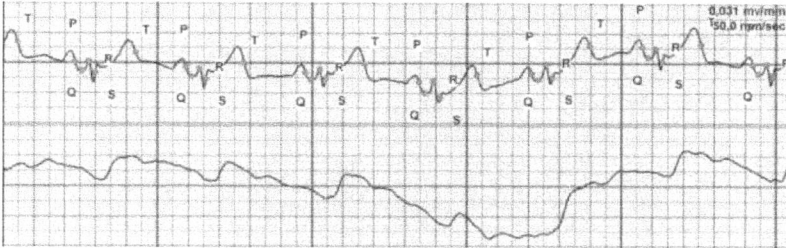

Diagnosis resulted from phase analysis.
The volume-related parameters of function of the heart and the associated blood vessels are within normal values.

Doctor's diagnosis.
The function of the aortic valve. The valve is opened with a delay. On the ECG, after the wave point S, there is a delay in the aortic pressure increase.
Anatomic peculiarities of aortic valve in the tension phase indicated on the ECG curve – norm.
Anatomic peculiarities of the aortic valve in the full opening phase indicated on the ECG curve. A changed condition indicated as smoothing of the wave Lj is available.
Flabbiness of aorta – not available.
Condition of elasticity of aorta – considerably reduced.
The apex of the rheogram is bifurcated, and the second apex goes far beyond the wave point T on the ECG curve.
Condition of contraction function of ventricular septum – the function is considerably diminished.
The R apex amplitude is less than that of the wave T.
Condition of contraction function of myocardium – the function is significantly diminished.

The amplitude of the lower portion of the QRS complex is a minimum, and that indicates the maximum passivity of the myocardium contraction. The QRS complex has a very low amplitude, and that is an indication of weakness of the heart as a whole (s. phase analysis).

Condition of function of the lungs – the function of the lungs is significantly diminished. A minor modulation of the rheogram curve by breathing is available.

Potential stroke problem – no sudden hemopulses are available. Potential stroke problem is not available.

ORTHOSTATIC TEST WAS NOT PERFORMED.

Additional information. Due to a delay in the valve opening, a pressure difference in the aorta does not reach normal values and remains small at a level of 20 mm Hg that is evident from the relevant RHEO.

SV (ml) – Stroke volume
SV(ml)=56.86 %=0.00   42.49   97.12

MV (l) – Minute volume
MV(l)=4.65 %=0.00   3.47   7.94

PV1 (ml) – Volume of blood in phase of rapid and slow filling of left (aortic) ventricle of heart
PV1(ml)=37.70 %=0.00   25.30   57.07

PV2 (ml) – Volume of blood flowing into left (aortic) ventricle of heart during systole
PV2(ml)=19.16 %=0.00   17.19   40.04

PV3 (ml) – Volume of blood ejected by left (aortic) ventricle of heart during rapid ejection phase
PV3(ml)=33.73 %=0.00   25.20   57.71

PV4 (ml) – Volume of blood ejected by left (aortic) ventricle of heart during slow ejection phase
PV4(ml)=23.13 %=0.00   17.29   39.41

PV5 (ml) – Volume of blood transferred by ascending aorta functioning as peristaltic pump
PV5(ml)=8.53 %=0.00   6.75   11.09

	QRS	RS	QT	PQ	TT	SV (ml)	MV (1)	PVl (ml)	PV2 (ml)	PV3 (ml)	PV4 (ml)	PV5 (ml)
Min	0.080	0.035	0.317	0.075	0.734	42.5	3.5	25.3	17.2	25.2	17.3	6.7
Max	0.110	0.060	0.317	0.075	0.734	97.1	7.9	57.1	40.0	57.7	39.4	11.1
Average	0.092	0.041	0.338	0.041	0.734	56.9	4.6	37.7	19.2	33.7	23.1	8.5
0	0.084	0.042	0.345	0.046	0.766	59.8	4.7	40.7	19.0	35.5	24.3	9.1
1	0.093	0.042	0.341	0.038	0.703	58.8	5.0	37.7	21.1	34.9	23.9	8.8
2	0.093	0.038	0.345	0.038	0.779	49.6	3.8	34.2	15.4	29.4	20.2	7.8
3	0.097	0.042	0.337	0.038	0.729	58.1	4.8	38.3	19.8	34.5	23.6	8.5
4	0.088	0.042	0.328	0.046	0.729	58.1	4.8	38.2	19.9	34.5	23.6	8.5
5	0.093	0.042	0.341	0.038	0.720	58.8	4.9	38.6	20.1	34.9	23.9	8.8
6	0.097	0.042	0.337	0.034	0.708	58.1	4.9	38.0	20.1	34.5	23.6	8.5
7	0.088	0.038	0.341	0.042	0.766	49.6	3.9	33.7	15.9	29.4	20.2	7.8
8	0.097	0.042	0.345	0.038	0.737	58.8	4.8	38.8	20.0	34.9	23.9	8.8
9	0.088	0.042	0.324	0.051	0.708	57.7	4.9	36.4	21.3	34.3	23.5	8.4
10	0.093	0.042	0.337	0.038	0.733	58.4	4.8	39.2	19.2	34.7	23.8	8.7

Comparative analysis of example No. 5 from practice

Phase duration.

The QRS complexes are equal in their duration.
A difference in the phase duration between the PQ phases is 0.099s.
The QT phases are equal in their duration.

Diagnoses.

Classical diagnosis option: hypertrophy of the right atrium. Typical signs of scarry changes within the anteroseptal area. Diffuse changes in myocardium.
The phase analysis shows the following.
Special attention should be paid to the small amplitude of the QRS complex. This is an indication of a weakness of the heart, and this fact is supported by the diagnosis produced automatically. All he-modynamic values are close to their lower limit of the norm – hypo-function. The small amplitudes of the QRS complex are indicative of stroke. But there is no stroke determined in this case. The wave T has a great amplitude. This is indicative of the compensation function maintaining normal he-modynamic parameters. Noted should be a flat apex on the RHEO curve that indicates a low elasticity of the aorta.
The therapy target should be an improvement in the cerebral circulation since it is the main cause of the processes responsible for reduction in the QRS complex amplitudes. In this connection, it should be mentioned that for this purpose additional tomographic testing & examinations are required.

Example No. 6 from Practice

Antecedent Anamnesis
Patient (full name: XX6), male, born 1965. Since 1996 registered as pulmonary fibrous cavernous tuberculosis patient.

Diagnosis
Pulmonary fibrous cavernous tuberculosis, infiltration & seeding phase, IIB, TMB(+).

Clinical Examination
Complaints: dyspnea and weakness. Basic data: weight – 53 kgs, stature – 176 cm; body mass index (BMI) – 16.91 kg/m2 (body mass deficit).

Dynamic data: arterial pressure (AP) 10 0/70, 9 0/6 0.

Heart frequency: 86 beats per minute, breathing frequency: 21–22 breath movements per minute.

Auscultation: heart tones are clear and rhythmic. Breathing in lungs: weakened.

Abdomen: soft, no pains available. Liver borders: the liver borders are extended (the liver is palpated extending 3cm over the edge of the costal margin). Urination is without pains, Pasternatsky's symptom is negative.

Tests
Laboratory Testing.

General blood test: leukocytes – 7.0*10*9, erythrocytes – 2.52*10*12, hemoglobin 92 g/l; cyclic immunocomplex – 1, erythrocyte sedimentation rate (ESR) – 60 mm/h; glucose – 4.71; bilirubin, total – 11.1 μmole/l, bilirubin conjugated – 0.0; thymol test 39; AST – 25 μmole/l.h, ALT – 0.67 μmole/l.h, C-reactive protein – negative.

Bacterial inoculation of sputum: negative. Bacterial inoculation of urine: negative. Urine analysis: no pathology available.

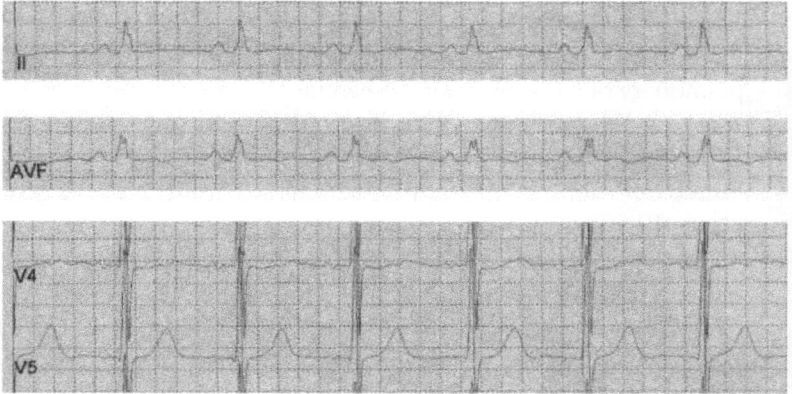

Instrumentation testing. ECG: sinus rhythm. Myocardial hypertrophy of the right atrium. Myocardial hypertrophy of the right ventricle of S-type. Re-polarization disturbances in the bottom wall of the left ventricle, possible of type of subepiocardiac insufficiency.

Radiographic examination:

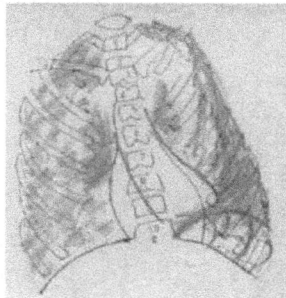

Right side: there is a nidal & focal infiltration throughout the lung available. Left side: the lung volume is decreased, there are massive pleural thickenings are available.

Date of entry – 05.04.2005
Date of birth – 28.06.1965
Full name – XX6 05-04-05 09-25-57 am

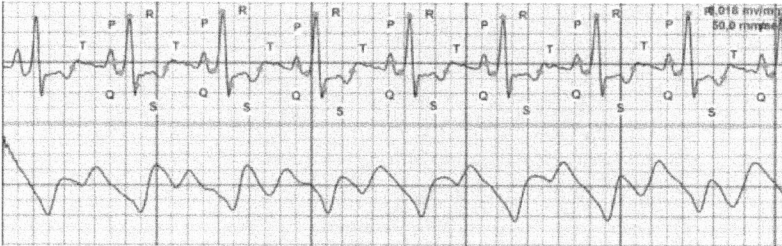

Diagnosis resulted from phase analysis.
The volume-related parameters of function of the heart and the associated blood vessels are within normal values.

Doctor's diagnosis.
The function of the aortic valve – norm.
Anatomic peculiarities of aortic valve in the tension phase indicated on the ECG curve. A changed condition as the smoothed SL wave is available.
Anatomic peculiarities of the aortic valve in the full opening phase indicated on the ECG curve. A changed condition indicated as smoothing of the wave Lj is available.
Flabbiness of aorta – apparently available. But a decrease in the aortic pressure, according to the RHEO curve, is starting after all near-normal values are reached.
Condition of elasticity of aorta – norm.
The apex of the rheogram is left referred to the wave point T on the ECG curve.
Condition of contraction function of ventricular septum – norm.
Condition of contraction function of myocardium – the function is significantly diminished. The amplitude of the lower portion of the

291

QRS complex is a minimum, and that indicates the maximum passivity of the myocardium contraction.

Condition of venous blood flow – the venous blood flow is heavily hindered.

There is an increase in the pressure, according to the RHEO, after dicrotic valley, available,

and the said increase exceeds the maximum pressure which is available in the systole.

Condition of function of the lungs – norm.

Potential stroke problem – no sudden hemopulses are available. Potential stroke problem is not available.

ORTHOSTATIC TEST WAS NOT PERFORMED.

SV (ml) – Stroke volume
SV(ml)=66.95 X=0.00   40.55   90.32

MV (l) – Minute volume
MV(l)=6.73 X=0.00   4.08   9.08

PV1 (ml) – Volume of blood in phase of rapid and slow filling of left (aortic) ventricle of heart
PV1(ml)=35.34 X=0.00   19.62   42.89

PV2 (ml) – Volume of blood flowing into left (aortic) ventricle of heart during systole
PV2(ml)=31.61 X=0.00   20.93   47.42

PV3 (ml) – Volume of blood ejected by left (aortic) ventricle of heart during rapid ejection phase
PV3(ml)=39.77 X=0.00   24.06   53.70

PV4 (ml) – Volume of blood ejected by left (aortic) ventricle of heart during slow ejection phase
PV4(ml)=27.18 X=0.00   16.49   36.82

PV5 (ml) – Volume of blood transferred by ascending
aorta functioning as peristaltic pump
PV5(ml)=8.13 X=0.00   6.04   9.42

	QRS	RS	QT	PQ	TT	SV (ml)	MV (1)	PVI (ml)	PV2 (ml)	PV3 (ml)	PV4 (ml)	PV5 (ml)
Min	0.080	0.035	0.286	0.064	0.597	40.5	4.1	19.6	20.9	24.1	16.5	6.0
Max	0.110	0.060	0.286	0.064	0.597	90.3	9.1	42.9	47.4	53.7	36.6	9.4
Average	0.107	0.049	0.295	0.043	0.597	67.0	6.7	35.3	31.6	39.8	27.2	8.1
0	0.105	0.046	0.290	0.042	0.602	61.5	6.1	33.5	27.9	36.5	25.0	7.6
1	0.109	0.051	0.299	0.046	0.610	71.3	7.0	37.7	33.5	42.3	28.9	8.5
2	0.109	0.051	0.299	0.046	0.602	71.3	7.1	36.7	34.6	42.3	28.9	8.5
3	0.109	0.051	0.295	0.046	0.602	70.6	7.0	36.9	33.7	41.9	28.7	8.3
4	0.105	0.046	0.295	0.046	0.606	62.0	6.1	32.9	29.2	36.8	25.2	7.7
5	0.109	0.051	0.303	0.046	0.610	71.9	7.1	37.6	34.4	42.7	29.2	8.7
6	0.109	0.046	0.290	0.046	0.581	60.9	6.3	29.4	31.5	36.2	24.7	7.4
7	0.109	0.051	0.295	0.042	0.606	70.6	7.0	38.5	32.1	41.9	28.7	8.3
8	0.105	0.051	0.295	0.042	0.602	71.3	7.1	39.0	32.3	42.3	28.9	8.5
9	0.105	0.051	0.295	0.042	0.598	71.3	7.2	38.5	32.8	42.3	28.9	8.5
10	0.105	0.046	0.299	0.042	0.598	62.6	6.3	32.8	29.8	37.2	25.4	7.9
11	0.105	0.046	0.295	0.042	0.598	62.0	6.2	33.0	29.1	36.8	25.2	7.7
12	0.118	0.055	0.303	0.038	0.598	80.1	8.0	42.0	38.0	47.6	32.5	9.0
13	0.105	0.051	0.295	0.042	0.589	71.3	7.3	37.4	33.9	42.3	28.9	8.5
14	0.105	0.051	0.290	0.042	0.589	70.6	7.2	37.6	33.0	41.9	28.7	8.3
15	0.105	0.046	0.295	0.042	0.593	62.0	6.3	32.5	29.5	36.8	25.2	7.7
16	0.109	0.051	0.303	0.038	0.589	71.9	7.3	37.2	34.7	42.7	29.2	8.7
17	0.105	0.051	0.282	0.038	0.568	69.2	7.3	36.4	32.8	41.1	28.1	8.0
18	0.097	0.046	0.290	0.051	0.602	62.6	6.2	33.1	29.4	37.2	25.4	7.9
19	0.105	0.046	0.295	0.042	0.585	62.0	6.4	31.5	30.6	36.8	25.2	7.7
20	0.105	0.051	0.295	0.042	0.593	71.3	7.2	37.9	33.3	42.3	28.9	8.5
21	0.109	0.051	0.290	0.042	0.593	69.9	7.1	37.2	32.8	41.5	28.4	8.1
22	0.105	0.042	0.295	0.042	0.602	53.2	5.3	28.2	25.0	31.6	21.6	7.0
23	0.114	0.051	0.299	0.038	0.598	70.6	7.1	37.6	33.0	41.9	28.7	8.3
24	0.105	0.046	0.299	0.042	0.606	62.6	6.2	33.7	28.9	37.2	25.4	7.9
25	0.101	0.046	0.295	0.042	0.598	62.6	6.3	33.8	28.8	37.2	25.4	7.9
26	0.105	0.046	0.295	0.042	0.593	62.0	6.3	32.5	29.5	36.8	25.2	7.7

Date of entry – 30.05.2005
Date of birth – 28.06.1965
Full name – XX6 30-05-05 12-18-24 pm

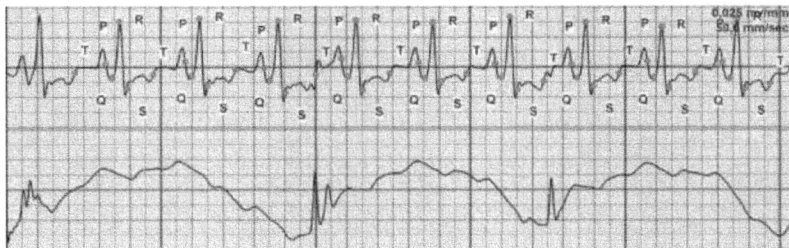

Diagnosis resulted from phase analysis.
The volume-related parameters of function of the heart and the associated blood vessels are within normal values.

Doctor's diagnosis.
The function of the aortic valve – norm. Anatomic peculiarities of aortic valve in the tension phase indicated on the ECG curve. The function is within the norm. But there is a change in anatomy of the valve available. The wave S has a small amplitude.

Anatomic peculiarities of the aortic valve in the full opening phase indicated on the ECG curve. A changed condition indicated as smoothing of the wave Lj is available. There is a very small increase in the aortic pressure available in this phase, according to the RHEO. And in the next phase, there is practically no pumping function of the aorta detectable. This is a significant pathology.

Flabbiness of aorta – not available.

Condition of elasticity of aorta – slightly reduced. The apex of the rheogram is flat.

Condition of contraction function of ventricular septum – norm.

Condition of contraction function of myocardium – the function is significantly diminished. The amplitude of the lower portion of the

QRS complex is a minimum, and that indicates the maximum passivity of the myocardium contraction.

Condition of venous blood flow – the venous blood flow is heavily hindered.

There is an increase in the pressure, according to the RHEO, after dicrotic valley available,

and the said increase exceeds the maximum pressure which is available in the systole.

Condition of function of the lungs – the function of the lungs is considerably diminished. A heavy modulation of the rheogram by breathing is available.

Potential stroke problem – apical hemopulses are available. An apex is similar to the wave R on the ECG curve.

ORTHOSTATIC TEST WAS NOT PERFORMED.

Additional information. According to the ECG, the wave R is lowered, and that indicates a pathology. Despite the fact that the systolic pressure is low and that there are also other problems available, the hemodynamic parameters are in this case within the norm (s. phase analysis).

SV (ml) – Stroke volume

SV(ml)=60.67 %=0.00	38.80		84.20

MV (l) – Minute volume

MV(l)=7.28 %=0.00	4.66		10.11

PV1 (ml) – Volume of blood in phase of rapid and slow filling of left (aortic) ventricle of heart

PV1(ml)=19.58 %=0.00	11.69		24.68

PV2 (ml) – Volume of blood flowing into left (aortic) ventricle of heart during systole

PV2(ml)=41.09 %=0.00	27.11		59.52

PV3 (ml) – Volume of blood ejected by left (aortic) ventricle of heart during rapid ejection phase

PV3(ml)=36.04 %=0.00	23.03		50.09

PV4 (ml) – Volume of blood ejected by left (aortic) ventricle of heart during slow ejection phase

PV4(ml)=24.63 %=0.00	15.78		34.11

PV5 (ml) – Volume of blood transferred by ascending
aorta functioning as peristaltic pump

PV5(ml)=7.24 %=0.00	5.45		8.06

	QRS	RS	QT	PQ	TT	SV (ml)	MV (1)	PV1 (ml)	PV2 (ml)	PV3 (ml)	PV4 (ml)	PV5 (ml)
Min	0.080	0.035	0.262	0.055	0.500	38.8	4.7	11.7	27.1	23.0	15.8	5.5
Max	0.110	0.060	0.262	0.055	0.500	84.2	10.1	24.7	59.5	50.1	34.1	8.1
Average	0.100	0.047	0.275	0.040	0.500	60.7	7.3	19.6	41.1	36.0	24.6	7.2
0	0.097	0.046	0.270	0.038	0.510	59.8	7.0	24.4	35.4	35.5	24.3	7.1
1	0.097	0.046	0.282	0.042	0.531	61.6	7.0	25.4	36.1	36.6	25.0	7.6
2	0.101	0.046	0.299	0.038	0.522	63.2	7.3	20.6	42.6	37.5	25.7	8.1
3	0.097	0.042	0.274	0.042	0.480	51.9	6.5	9.8	42.1	30.8	21.1	6.6
4	0.105	0.051	0.270	0.038	0.493	67.1	8.2	21.9	45.2	39.9	27.2	7.4
5	0.101	0.046	0.274	0.038	0.501	59.8	7.2	20.7	39.1	35.5	24.3	7.1
6	0.101	0.046	0.270	0.038	0.489	59.2	7.3	18.0	41.2	35.2	24.0	7.0
7	0.097	0.046	0.274	0.042	0.497	60.4	7.3	18.3	42.1	35.9	24.5	7.3
8	0.101	0.046	0.265	0.042	0.480	58.6	7.3	14.5	44.1	34.8	23.8	6.8
9	0.105	0.051	0.270	0.042	0.489	67.1	8.2	18.3	48.8	39.9	27.2	7.4
10	0.097	0.042	0.282	0.038	0.501	52.8	6.3	16.1	36.8	31.4	21.5	6.8
11	0.105	0.051	0.265	0.038	0.476	66.4	8.4	16.9	49.5	39.4	26.9	7.2
12	0.097	0.046	0.270	0.046	0.505	59.8	7.1	20.1	39.7	35.5	24.3	7.1
13	0.101	0.046	0.278	0.038	0.510	60.4	7.1	22.1	38.3	35.9	24.5	7.3
14	0.097	0.046	0.278	0.042	0.514	61.0	7.1	22.2	38.8	36.2	24.8	7.4

Comparative analysis of example No. 6 from practice

Diagnoses.

Classical diagnosis option: hypertrophy of the right atrium and ventricle.

The phase analysis shows the following.

In this case, there are data obtained from two examinations of the patient which were carried out with the analyzer CAR-DIOCODE. The said examinations were performed with an interval approximately of two months.

Of interest should be two factors. The first factor is the following: there is an inversion of the wave T on the ECG curve available. This is a classical inversion of the wave T because the slow ejection phase is not affected. In this case, the upper portion of the QRS complex is within the norm, and the lower portion indicates a decrease in the contraction function of the ventricles. The wave P is also within the norm. The hemodynamic parameters are within the norm.

The second factor of interest is an increase in the arterial pressure on the RHEO curve after the dicrotic valley. This is an indication of the complete disturbance in venous flow.

The second examination performed two months later gives us changes in the general picture. The contraction function of the ventricles is diminished, and the amplitude of the wave S is decreased. The wave P is increased. The point Q on the ECG curve is positioned lower considering the coordinate axis. This is a very bad symptom since this phase being analyzed anatomically is determined by the heart area which is the basement of the valve fixtures. The drop of the wave point Q is an indication of the "total failure" of the mechanical basement of the valve system. The inverted wave T is not expressed symptomatically. But during the second examination detected are the AP hemopulses that are signals of potential stroke development. They are apical, but not dangerous. More important is the fact that the modulation of the RHEO by breathing becomes heavier. The modulation is in this case significantly expressed. This indicates a considerable diminishing of the lung function.

The above mentioned patient died nine months later. Death cause: cardiopulmonary decompensation.

Example No. 7 from Practice

Antecedent Anamnesis
Patient (full name: XX7), female, born 1959. Since 2001 registered as pulmonary tuberculous patient.

Diagnosis
Disseminated pulmonary tuberculosis, infiltration & destruction phase, IA, TMB+. Essential hypertension, degree I. Alcoholic hepatitis.

Clinical Examination
Complaints: dyspnea, cough accompanied by grey sputum, hyperhidrosis, pectoral pains, pains in stomach. Basic data: weight – 57 kgs, stature – 165 cm; body mass index (BMI) – 17. 9 4 k g / m 2 (body mass deficit). Dynamic data: arterial pressure (AP) 13 0/9 0 , 14 0/9 0 , 16 0/9 0. Heart frequency: 96 beats per minute, breathing frequency: 22–24 breath movements per minute.

Auscultation: heart tones are damped down and rhythmic. Breathing in lungs: dry and moist rales on both sides. Abdomen: pains in epigastrium and the right hypochondrium are available. Liver orders: the liver border extends 3 cm beyond the edge of the costal margin.

Tests
Laboratory Testing.
General blood test: leukocytes – 8.6*10*9, erythrocytes – 4.22*10*12, hemoglobin 112 g/l, erythrocyte sedimentation rate (ESR) – 58 mm/h, prothrombin – 106%, glucose –3.3 μmole/l; bilirubin, total – 8.6, bilirubin conjugated – 0.0; AST – 0.25 μmole/l.h, ALT – 0.67 μmole/l.h. Bacterial inoculation of sputum – TMB 5 c. Urine analysis: no pathology available.

Instrumentation testing.
ECG: P – 0.10; PQ – 0.12; QRS – 0.09;
QRST – 0.39; R–R – 0.64.
Heart frequency – 94 beats per minute.
Moderate sinus tachycardia. Myocardial hypertrophy of the right atrium. Diffuse changes in myocardium.

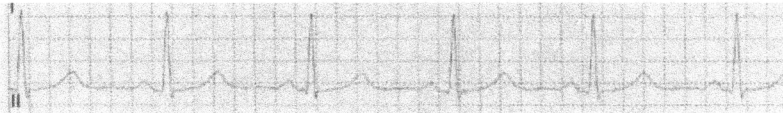

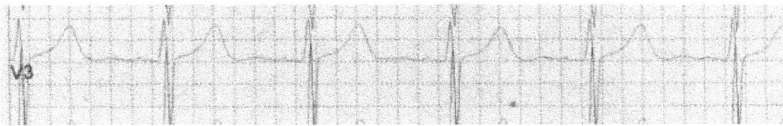

Radiographic examination:

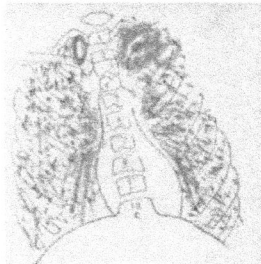

On both sides, throughout the total length: a small-scale nidal dissemination of discharge type, in S1 ? SII from both sides -destruction cavities with infiltrated walls; left side in S1: a focus of 5*4 cm with destruction available.

Date of entry – 03.06.2005
Date of birth – 13.08.1959
Full name – XX7 03-06-05 09-42-09 am

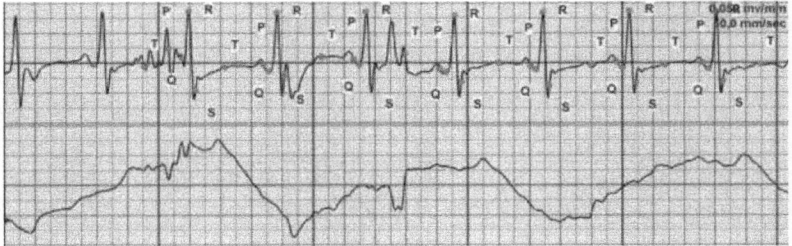

Diagnosis resulted from phase analysis.
The volume-related parameters of function of the heart and the associated blood vessels are within normal values.

Doctor's diagnosis.
The function of the aortic valve – there is a delay in opening the
valve available.
Anatomic peculiarities of aortic valve in the tension phase indicated on the ECG curve – norm.
Anatomic peculiarities of the aortic valve in the full opening phase
indicated on the ECG curve. A changed condition indicated as
smoothing of the wave Lj is available. But the pressure does not
reach its normal value.
Condition of elasticity of aorta – norm.
The apex of the rheogram is pyramidal and positioned left referred
to the wave point T on
the ECG curve.
Condition of contraction function of ventricular septum – norm.
Condition of contraction function of myocardium – the function is
weakened. The amplitude of the lower portion of the QRS complex
is a half of the norm.
Condition of venous blood flow – norm.

Condition of function of the lungs – the function of the lungs is considerably diminished. A heavy modulation of the rheogram by breathing is available.
Potential stroke problem – sudden hemopulses are available. Potential stroke problem is available.

ORTHOSTATIC TEST WAS NOT PERFORMED.

SV (ml) – Stroke volume
SV(ml)=59.28 %=0.00   41.66   94.20

MV (l) – Minute volume
MV(l)=6.20 %=0.00   4.36   9.85

PV1 (ml) – Volume of blood in phase of rapid and slow filling of left (aortic) ventricle of heart
PV1(ml)=27.74 %=0.00   16.22   35.86

PV2 (ml) – Volume of blood flowing into left (aortic) ventricle of heart during systole
PV2(ml)=31.53 %=0.00   25.44   58.34

PV3 (ml) – Volume of blood ejected by left (aortic) ventricle of heart during rapid ejection phase
PV3(ml)=35.18 %=0.00   24.71   55.99

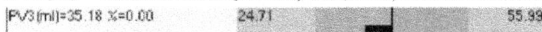

PV4 (ml) – Volume of blood ejected by left (aortic) ventricle of heart during slow ejection phase
PV4(ml)=24.10 %=0.00   16.95   38.21

PV5 (ml) – Volume of blood transferred by ascending
aorta functioning as peristaltic pump
PV5(ml)=8.25 %=0.00   6.44   10.35

	QRS	RS	QT	PQ	TT	SV (ml)	MV (1)	PVl (ml)	PV2 (ml)	PV3 (ml)	PV4 (ml)	PV5 (ml)
Min	0.080	0.035	0.303	0.062	0.574	41.7	4.4	16.2	25.4	24.7	16.9	6.4
Max	0.110	0.060	0.303	0.062	0.574	94.2	9.8	35.9	58.3	56.0	38.2	10.4
Average	0.092	0.043	0.313	0.040	0.574	59.3	6.2	27.7	31.5	35.2	24.1	8.2
0	0.084	0.042	0.273	0.088	0.517	53.1	6.2	4.8	48.3	31.5	21.6	7.0
1	0.097	0.046	0.336	0.038	0.626	67.7	6.5	36.7	30.9	40.1	27.5	9.5
2	0.088	0.042	0.298	0.034	0.542	55.2	6.1	25.3	29.9	32.8	22.4	7.6
3	0.097	0.042	0.340	0.038	0.605	58.2	5.8	27.8	30.4	34.5	23.7	8.6
4	0.088	0.042	0.353	0.034	0.592	59.8	6.1	26.5	33.3	35.5	24.3	9.2
5	0.088	0.042	0.277	0.038	0.483	53.1	6.6	12.6	40.5	31.5	21.6	7.0
6	0.092	0.042	0.332	0.038	0.613	57.8	5.7	30.4	27.4	34.3	23.5	8.5
7	0.097	0.046	0.319	0.038	0.576	65.9	6.9	30.5	35.4	39.1	26.8	8.9
8	0.097	0.042	0.290	0.034	0.525	53.5	6.1	21.8	31.7	31.8	21.7	7.1
9	0.092	0.042	0.231	0.038	0.504	46.7	5.6	23.7	23.0	27.8	19.0	5.2
10	0.092	0.046	0.315	0.034	0.660	65.9	6.0	42.7	23.2	39.1	26.8	8.9
11	0.092	0.046	0.387	0.034	0.643	72.6	6.8	36.1	36.6	43.1	29.5	11.2

Comparative analysis of example No. 7 from practice

Phase duration.

The QRS complexes are equal in their duration.
A difference in the duration of the PQ phases is 0.08 s.

Diagnoses.

Classical diagnosis option: Hypertrophy of myocardium of the right atrium. Diffuse changes in myocardium.
The phase analysis shows the following.
Special attention should be paid to the artifacts on the ECG curve. They represent spontaneous excitations of unknown nature. They are accompanied by formation of the hemopulses of the AP registered synchronously on the ECG curve. It should be noted that the AP hemopulses are of negative values. The RHEO shows sudden pulse drops of the AP. One among them is of rectangular shape. This shape is an indication of a danger. An increase in the pulse length indicates a stroke. A negative pulse leads to cerebral infarction, and a positive pulse leads to hemorrhagic stroke.
Of interest is a heavy modulation of the RHEO by breathing waves. It is an indication of lung weakness. In this case, the modulation is significant. The wave T is not available. But hemodynamic parameters are within the norm.
The therapy target should be an improvement in the function of the lungs and blood vessels.

Example No. 8 from Practice

Antecedent Anamnesis
Patient (full name: XX8), male, born 1952. 1999 right-side pneumothorax registered. 2001–2003 the patient registered as pulmonary tuberculous patient. Infiltrative tuberculosis of the right lung.

Diagnosis
Chronic obstructive disease of lungs, chronic cardiopulmonary decompensation, degree II, III. Pulmonary tuberculosis clinically healed.

Clinical Examination
Complaints: dyspnea at a minor physical activity and weakness, cough accompanied by yellow sputum, edema of feet and legs.
Basic data: weight – 57 kgs, stature – 165 cm; body mass index (BMI) – 20.73 kg/m2, 21 kg/m2 (normal body mass). Acrocyanosis. Dynamic data: arterial pressure (AP) 130/90, 120/80. Heart frequency: 80 beats per minute, breathing frequency: 19–20 breath movements per minute.
Auscultation: heart tones are damped down and rhythmic.
Breathing in lungs: breathing weakened on both sides, random moist rales on both sides. Abdomen: swollen, pains in the right hypochondrium are available. Liver borders: the liver border extends 5 cm beyond the edge of the costal margin. Urination: no deviations.

Tests
Laboratory Testing.
General blood test: leukocytes – 8.8*10*9, erythrocytes – 4.52*10*12, hemoglobin 116 g/l, erythrocyte sedimentation rate (ESR) – 2 mm/h, pro-thrombin – 106%, glucose –4.2; bi-lirubin, total – 10.3 μmole/l, bilirubin conjugated – 0.0; AST – 1,17 μmole/l.h,

ALT – 1.79 μmole/l.h. Bacterial inoculation – no growth detected. Urine analysis: protein 0.033, individual leukocytes in visual field.

Instrumentation testing. ECG: PQ – ff; QRS – 0.15; QRST – 0.32; R–R – 0.43;
Heart frequency – 139 beats per minute. Atrial fibrillation with regular contraction of ventricles is possibly caused by tachycardia of atrioventricular nature. Complete right bundle-branch block, left anterior fascicular block. Apparent changes in myocardium.

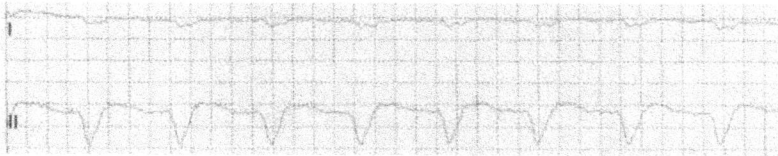

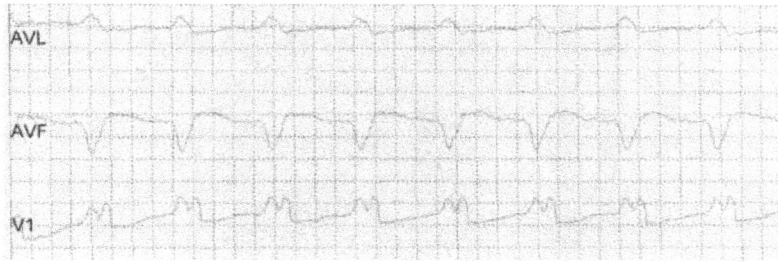

Radiographic examination:

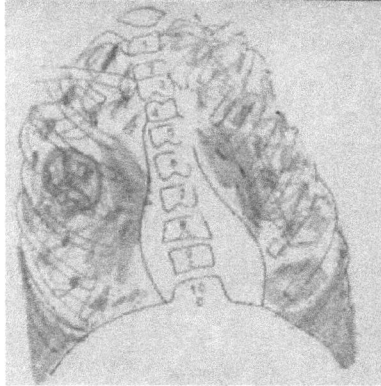

On both sides, throughout the length: ring fibrosis, nidi without clear contours are available; nidi are available on the right side, in intercostal area III, fusing in a focus of 4,0 cm with destruction; there is a non-clear cavity on the left side in the lower lobe available; there is exudates in sini available; the roots are expanded and poorly structured.

Date of entry – 14.06.2005
Date of birth – 03.03.1952
Full name – XX8 14-06-05 11-19-53 am

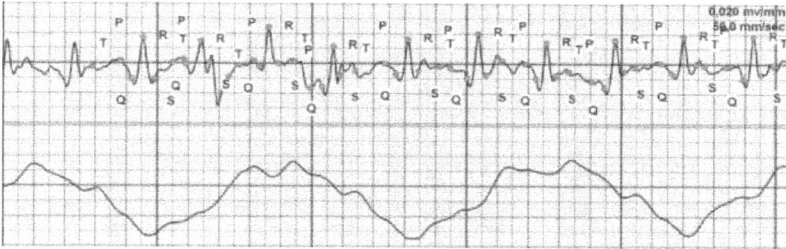

Diagnosis resulted from phase analysis.
Automatically generated analysis.
PV1 (l) – A dangerous abnormal deficit in the volume of blood entering the left ventricle. PV2 (l) - within the norm. PV5 (l) - within the norm.

Doctor's diagnosis.
The function of the aortic valve – norm.
Anatomic peculiarities of aortic valve in the tension phase indicated on the ECG curve. A changed condition as a smoothed wave SL is available. Blood enters the aorta at the beginning of the phase.
Anatomic peculiarities of the aortic valve in the full opening phase indicated on the ECG curve. A changed condition indicated as smoothing of the wave Lj is available. Actually, it cannot be detected. An increase in pressure in the aorta is very small, according to the RHEO.

Flabbiness of aorta –not available.

Condition of elasticity of aorta – norm.

The apex of the rheogram is parabolic and positioned left referred to the wave point T on the ECG curve.

Condition of contraction function of ventricular septum – norm.

Condition of contraction function of myocardium – the function is considerably diminished. The amplitude of the lower portion of the QRS complex is a minimum, and that is an indication of the maximum passivity of myocardium contraction.

Condition of venous blood flow. – Norm.

Condition of function of the lungs – the function of the lungs is considerably diminished. A heavy modulation of the rheogram by breathing is available.

Potential stroke problem – no sudden hemopulses are available. Potential stroke problem is not available.

ORTHOSTATIC TEST WAS NOT PERFORMED.

Additional information. There is a significant pathology of myocardium available. The wave S is not available. The P and T waves are not available, too. This creates a dangerous underloading on the ventricles by atrii (s. phase analysis). A mean deficit is 360% of the norm that is very dangerous.

SV (ml) – Stroke volume

| SV(ml)=69.82 %=0.00 | 37.66 | | 80.18 |

MV (l) – Minute volume

| MV(l)=9.38 %=0.00 | 5.06 | | 10.77 |

PV1 (ml) – Volume of blood in phase of rapid and slow filling of left (aortic) ventricle of heart

| PV1(ml)=-6.87 %=360.21 | 2.64 | | 5.41 |

PV2 (ml) – Volume of blood flowing into left (aortic) ventricle of heart during systole

| PV2(ml)=76.70 %=2.58 | 35.02 | | 74.77 |

PV3 (ml) – Volume of blood ejected by left (aortic) ventricle of heart during rapid ejection phase

| PV3(ml)=41.61 %=0.00 | 22.35 | | 47.71 |

PV4 (ml) – Volume of blood ejected by left (aortic) ventricle of heart during slow ejection phase

| PV4(ml)=28.22 %=0.00 | 15.30 | | 32.46 |

PV5 (ml) – Volume of blood transferred by ascending
aorta functioning as peristaltic pump

| PV5(ml)=5.08 %=0.09 | 5.09 | | 7.23 |

	QRS	RS	QT	PQ	TT	SV (ml)	MV (l)	PV1 (ml)	PV2 (ml)	PV3 (ml)	PV4 (ml)	PV5 (ml)	HR
Min	0.080	0.035	0.247	0.049	0.447	37.7	5.1	2.6	35.0	22.4	15.3	5.1	134.3
Max	0.110	0.060	0.247	0.049	0.447	80.2	10.8	5.4	74.8	47.7	32.5	7.2	134.3
Average	0.135	0.061	0.235	0.077	0.447	69.8	9.4	-6.9	76.7	41.6	28.2	5.1	134.3
0	0.147	0.071	0.265	0.067	0.513	95.3	11.1	22.1	73.2	56.8	38.5	7.0	117.0
1	0.151	0.059	0.273	0.017	0.362	73.3	12.2	783.5	-710.2	43.6	29.7	6.2	166.0
2	0.126	0.063	0.227	0.034	0.425	73.0	10.3	13.7	59.3	43.5	29.5	5.2	141.3
3	0.109	0.038	0.219	0.076	0.399	35.7	5.4	-23.4	59.0	21.2	14.5	3.7	150.2
4	0.143	0.067	0.282	0.071	0.542	95.5	10.6	25.2	70.3	56.8	38.6	8.1	110.6
5	0.126	0.050	0.231	0.105	0.404	54.0	8.0	-159.3	213.3	32.1	21.8	4.6	148.7
6	0.126	0.059	0.210	0.071	0.429	59.5	8.3	1.2	58.3	35.4	24.0	4.0	139.9
7	0.135	0.063	0.206	0.080	0.433	59.1	8.2	1.1	58.0	35.3	23.8	3.3	138.6
8	0.151	0.080	0.206	0.076	0.446	66.9	9.0	9.3	57.6	40.0	26.9	2.6	134.7
9	0.118	0.050	0.198	0.109	0.446	46.4	6.2	-2.4	48.8	27.7	18.8	3.3	134.7
10	0.130	0.059	0.227	0.071	0.471	64.5	8.2	12.2	52.3	38.4	26.1	4.7	127.4
11	0.122	0.050	0.202	0.080	0.412	46.4	6.8	-7.1	53.5	27.7	18.8	3.3	145.6
12	0.139	0.071	0.282	0.097	0.530	106.0	12.0	5.1	100.9	63.1	42.9	8.8	113.3
13	0.122	0.050	0.210	0.088	0.374	49.1	7.9	-120.7	169.9	29.3	19.9	3.7	160.4
14	0.156	0.055	0.219	0.080	0.441	44.9	6.1	-0.7	45.6	26.8	18.1	2.6	135.9
15	0.147	0.080	0.265	0.092	0.517	111.6	12.9	11.8	99.8	66.5	45.1	7.6	116.0
16	0.147	0.076	0.277	0.088	0.454	109.6	14.5	-136.7	246.2	65.3	44.3	8.3	132.2

Comparative analysis of example No. 8 from practice

Phase duration.

The QRS complexes are equal in their duration.

A difference in the duration of the PQ phases is 0.08 s.

A difference in the duration of the QT phases is 0.077 s.

Diagnoses.

Classical diagnosis option: Hypertrophy of myocardium of the right atrium. Diffuse changes in myocardium.

The phase analysis shows the following.

It is evident that the phase criteria are distorted. It is difficult to analyze the phases on the ECG curve. The waves T & P are not available. The phases of the valve function are practically flattened. There is a heavy modulation of the RHEO by breathing waves available. A mean deficit in blood entering the ventricles in the early diastole is great and amounts approximately to 360%.

The therapy recommendations should be prepared on the basis of the actual patient's symptomatology; the proposed therapy should be aimed at an improvement in the actual health state as a whole supporting the patient to a certain degree.

It should be noted that one month later the patient died. Cause of death: cardiopulmo-nary decompensation.

Example No. 9 from Practice

Antecedent Anamnesis

Patient (full name: XX9), male, born 1927. Since 2002 registered as pulmonary tuberculous patient.

Diagnosis
Nidal tuberculosis S1-S2 of both lungs, infiltration phase, IA.
Clinical Examination
Complaints: dyspnea at physical activity and weakness.
Basic data: weight – 57 kgs, stature – 165 cm; body mass index (BMI) – 20.73 kg/m2 (n o r-mal body mass).
Dynamic data: arterial pressure (AP) 110/70, 12 0/70 , 12 0/8 0.
Heart frequency: 80 beats per minute, breathing frequency: 19–21 breath movements per minute.
Auscultation: heart tones are clear and rhythmic.
Breathing in lungs: vesicular. Abdomen: soft, no pains are available.
Liver borders: within the norm.

Tests
Laboratory Testing.
General blood test: leukocytes – 4.8*10*9, erythrocytes – 4.52*10*12, hemoglobin 140 g/l, erythrocyte sedimentation rate (ESR) – 10 mm/h, prothrombin – 106%, glucose – 4.17; bilirubin, total – 5.3, bi-lirubin conjugated – 0.0; AST – 50 μmole/ l.h, ALT – 1.57 μmole/l.h. Bacterial inoculation – negative. Urine analysis: no pathology.
Instrumentation testing:
ECG: P-0.08; PQ – 0.13; QRS – 0.13;
QRST – 0.43; R-R-100;
Heart frequency – 60 beats per minute.

The sinus rhythm is available. Complete right bundle-branch block. Apparent changes in myocardium.

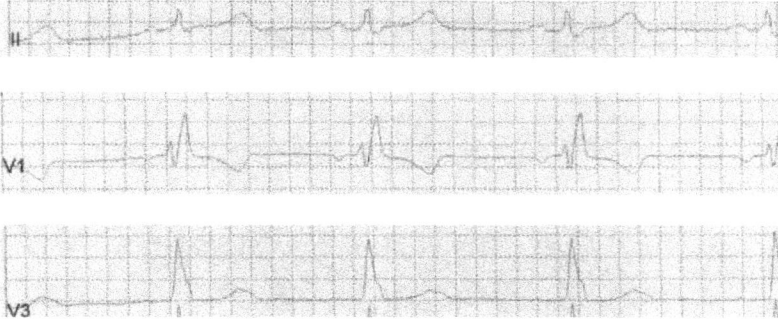

Radiographic examination:

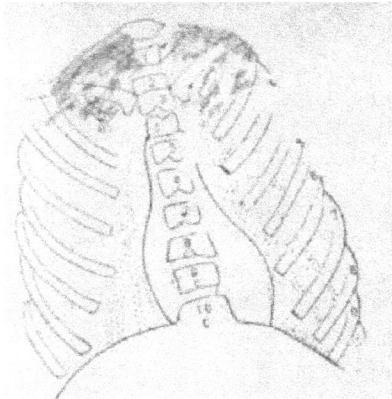

In S1–S2 of both lungs: there are apical thickenings available; nidi without clear contours are available; the roots are structured, the sini are free; the borders of the heart are within the norm.

315

Date of entry – 19.07.2005
Date of birth – 03.10.1927
Full name – XX9 19-07-2005 9-20-40

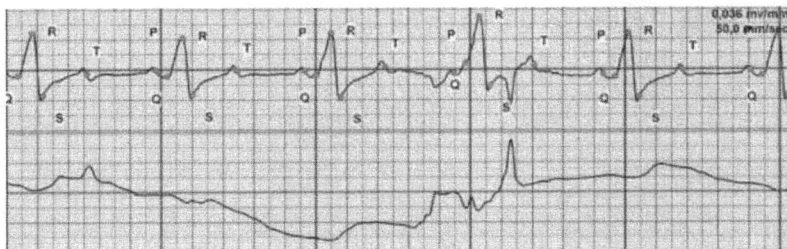

Diagnosis resulted from phase analysis.
Automatically generated analysis.

SV (ml.l) – An increased volume of blood enters the left heart. The heart is loaded by the blood volume. Possible is an increase in the blood pressure in lesser circulation. The parameter PV5 shall be thoroughly evaluated that is an indication of the blood volume pumped by the aorta as peristaltic pump.

MV1 (l) – An increased volume of blood enters the left heart. The heart is loaded by the blood volume. Possible is an increase in the blood pressure in lesser circulation. The parameter PV5 shall be thoroughly evaluated that is an indication of the blood volume pumped by the aorta as peristaltic pump. PV1 (l) – within the norm. PV2 (l) – There is an increase in loading by the blood volume on the left atrium in the atrial systole. Possible is an increase in the blood pressure in lesser circulation. A heavy loading by the volume is available (hyperfunction).

PV3 (l) – There is an increase in the blood volume ejected by the left ventricle in the rapid ejection phase. The arterial pressure in greater circulation should be closely studied. Possible is an increase in the arterial pressure.

PV4 (l) – There is an increase in the blood volume ejected by the left ventricle in the slow ejection phase. The arterial pressure in

316

greater circulation should be closely studied. Possible is an increase in the arterial pressure.

PV5 (l) – There is an increase in the blood volume pumped by the ascending aorta as peristaltic pump. The tonus of the ascending aorta is strengthened. The consequence is a decrease in loading on the left ventricle due to a reduced resistance by the aorta.

Doctor's diagnosis

The function of the aortic valve – the valve is opened with a delay. A delay in the aortic pressure increase is available on the ECG curve after wave point S.

Anatomic peculiarities of aortic valve in the tension phase indicated on the ECG curve. A changed condition as a smoothed wave SL is available.

Anatomic peculiarities of the aortic valve in the full opening phase indicated on the ECG curve. - A changed condition indicated as smoothing of the wave Lj is available. An increase in pressure is very small in this case. A difference in pressures between the maximum and the minimum, according to the RHEO, is small. The wave T can be poorly detected that indicates a weak pumping function of the aorta.

Flabbiness of aorta – not available.

Condition of elasticity of aorta – slightly diminished. The apex of the rheogram is flat.

Condition of contraction function of ventricular septum – there are some features of a pathological expansion of the QRS complex available. The duration of the QR phase is increased due to a reduction in the rate of increase of its amplitude. It indicates a slow contraction of the ventricular septum while its contraction amplitude is normal.

Condition of contraction function of myocardium – the function is considerably diminished. The amplitude of the lower portion of the

QRS complex is a minimum, and that is an indication of the maximum passivity of myocardium contraction.

Condition of venous blood flow – norm.

Condition of function of the lungs – the function of the lungs is diminished. A minor modulation of the rheogram by breathing is available.
Potential stroke problem – apical hemopulses are available. The apex is similar to the R wave point on an ECG curve.

ORTHOSTATIC TEST WAS NOT PERFORMED.

The hemodynamic parameters show that an increased tension (stress condition) in the function of the heart and the associated blood vessels are available (s. diagnosis based on the phase analysis).

SV (ml) – Stroke volume
SV(ml)=126.17 %=18.84   45.08   106.16

MV (l) – Minute volume
MV(l)=7.68 %=18.84   2.74   6.46

PV1 (ml) – Volume of blood in phase of rapid and slow filling of left (aortic) ventricle of heart
PV1(ml)=77.65 %=7.59   30.93   72.17

PV2 (ml) – Volume of blood flowing into left (aortic) ventricle of heart during systole
PV2(ml)=48.52 %=42.74   14.15   33.93

PV3 (ml) – Volume of blood ejected by left (aortic) ventricle of heart during rapid ejection phase
PV3(ml)=74.92 %=18.84   26.73   63.04

PV4 (ml) – Volume of blood ejected by left (aortic) ventricle of heart during slow ejection phase
PV4(ml)=51.25 %=18.85   18.35   43.13

PV5 (ml) – Volume of blood transferred by ascending
aorta functioning as peristaltic pump
PV5(ml)=16.22 %=13.43   7.77   13.58

318

	QRS	RS	QT	PQ	TT	SV (ml)	MV (1)	PV1 (ml)	PV2 (ml)	PV3 (ml)	PV4 (ml)	PV5 (ml)	HR
Min	0.080	0.035	0.367	0.087	0.985	45.1	2.7	30.9	14.2	26.7	18.4	7.8	60.9
Max	0.110	0.060	0.367	0.087	0.985	106.2	6.5	72.2	34.0	63.0	43.1	13.6	60.9
Average	0.162	0.065	0.444	0.075	0.985	126.2	7.7	77.6	48.5	74.9	51.3	16.2	60.9
0	0.160	0.067	0.441	0.084	1.055	132.1	7.5	84.3	47.9	78.5	53.7	16.7	56.9
1	0.164	0.071	0.446	0.084	1.038	145.6	8.4	91.6	54.0	86.5	59.1	17.8	57.8
2	0.164	0.071	0.441	0.076	1.021	144.7	8.5	92.1	52.6	85.9	58.7	17.6	58.7
3	0.156	0.063	0.437	0.071	0.988	119.1	7.2	75.0	44.0	70.7	48.4	15.5	60.7
4	0.160	0.067	0.441	0.067	0.992	132.1	8.0	84.2	47.9	78.5	53.7	16.7	60.5
5	0.160	0.067	0.441	0.080	0.983	132.1	8.1	80.8	51.3	78.5	53.7	16.7	61.0
6	0.164	0.063	0.450	0.071	0.971	119.8	7.4	72.7	47.1	71.1	48.7	15.8	61.8
7	0.164	0.063	0.450	0.067	0.958	119.8	7.5	72.7	47.0	71.1	48.7	15.8	62.6
8	0.156	0.067	0.441	0.076	0.962	132.9	8.3	81.3	51.5	78.9	54.0	16.9	62.3
9	0.156	0.067	0.437	0.076	0.962	132.1	8.2	81.2	50.9	78.5	53.7	16.7	62.3
10	0.164	0.059	0.446	0.076	0.958	106.5	6.7	62.9	43.6	63.2	43.3	14.4	62.6
11	0.164	0.063	0.446	0.067	0.954	119.1	7.5	72.3	46.8	70.7	48.4	15.5	62.9
12	0.164	0.063	0.450	0.071	0.958	119.8	7.5	71.8	47.9	71.1	48.7	15.8	62.6
13	0.164	0.067	0.450	0.080	0.992	132.9	8.0	80.8	52.1	78.9	54.0	16.9	60.5
14	0.168	0.059	0.450	0.080	0.988	106.5	6.5	63.2	43.2	63.2	43.3	14.4	60.7

Comparative analysis of example No. 9 from practice

Phase duration.

The QRS complexes differ in their duration (a difference is 0,032 s.).

A difference in the duration of the PQ phases is 0.055 s.

A difference in the duration of the QT phases is 0.014 s.

Diagnoses.

Classical diagnosis option: Complete right bundle-branch block. Apparent changes in myocardium.

The phase analysis shows the following.

Special attention should be paid to the expanded QRS complex and the flattened phases of the aortic valve function. They shall be considered as two important problems. What is an indication of the complete right bundle-branch block on the ECG curve registered with the analyzer CARDIOC-ODE? The phase QR is responsible for the contraction of the ventricular septum, and it is the His's bundle that controls this function. Changes of the ascending line from the wave Q to the apex R can be registered only on the ECG first-order derivative. Figure 110 shows that there is a bifurcation of the derivative apex. Normally, it should be a straight line. This bifurcation is a criterion that there is a disturbance in the conductivity of the ventricular septum. But we are not able to determine whether it is the right or left His' bundle responsible for this phenomenon. Considering the fact that there is a self-compensation of conductivity areas in the heart available, for this purpose, it is required to carry out additional tests and examinations. The pathology of the ventricular septum is registered by automatic calculation of the hemodynamic values. So, the parameter PV 2 characterizing the contraction function of the atrium is above the norm, and the detected disturbance in the conductivity produces an increased load on the ventricles. A deviation of 42% from the norm is a significant one. Other hemody-namic parameters exceed also their normal values. So, we can arrive at the conclusion that the heart has to function in this case under a heavy load.

The flattened phases of the valve function could be of the same nature like the above disturbance in the conductivity. Anatomically, they are close to each other.

There are arterial pressure hemopulses on the RHEO available. There is a combination formed by apical and rectangular pulses. It is an indication of a danger.

Weakness of the lung function is also detected. The criterion in this case is the RHEO modulation by breathing waves.

The therapy target should be an improvement in the symptomatology covering the heart area and occiput.

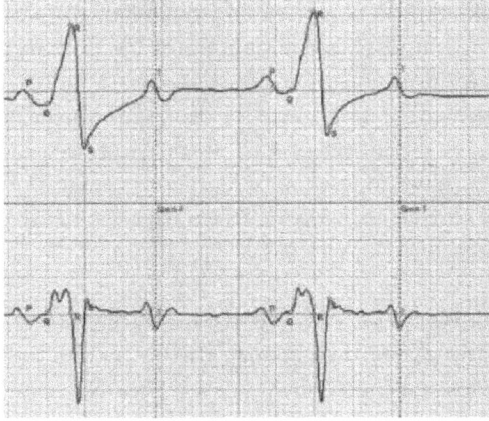

Fig. 110. Bifurcated Apex on the First-Order Derivative Curve in QR Phase.

Example No. 10 from Practice

Antecedent Anamnesis

Patient (full name: XX10), female, born 1926. Periodically registered complaints about hypertension.

Diagnosis

Coronary heart disease (CHD), atherosclerotic cardiosklerosis, Functional Class II, with periodic rhythm abnormalities. Essential hypertension, degree II. Pancreatic diabetes, type II Diabetes-induced retinopathy.

Clinical Examination

Complaints: dyspnea at physical activity and weakness.

Basic data: weight – 90 kgs, stature – 163 cm; body mass index (BMI) – 33.87 kg/m2 (obesity, degree II).

Dynamic data: arterial pressure (AP) 160/100, 140/ 80 , 120/80 . Heart frequency: 84, 90 beats per minute, breathing frequency: 18–19 breath movements per minute.

Auscultation: heart tones are damped down; accent of tone II on the aorta. Breathing in lungs: vesicular. Abdomen is soft, no pains available. The liver borders are within the norm.

Tests

Laboratory Testing.

General blood test: leukocytes – 4.8*10*9 (leukocyte formula shows no pathology), erythrocytes – 3.6*10*12, hemoglobin 123 g/l, erythrocyte sedimentation rate (ESR) – 3 mm/h, prothrombin – 106%, glucose – 10 μmole/l, cholesterin – 5.1 μmole/l, C-reactive protein – negative. Urine analysis: no pathology.

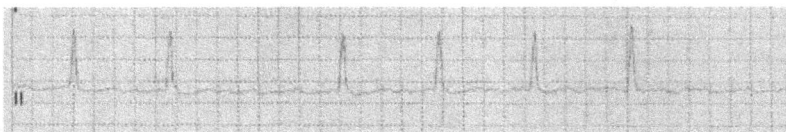

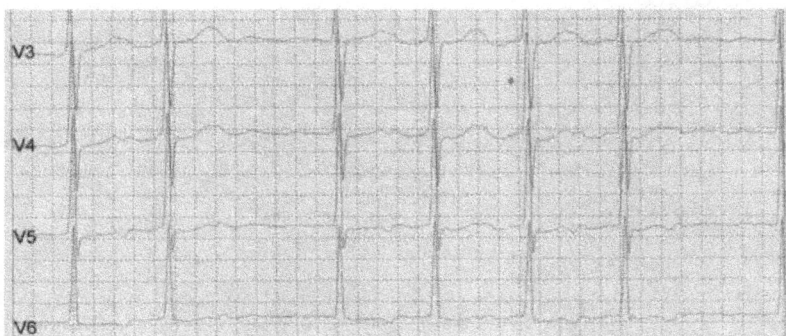

Instrumentation testing.
ECG: PQ – ff; QRS – 0.08; QRST – 0.36;
R–R – 0.80–0.45

Heart frequency – 75–133 beats per minute.

Atrial fibrillation with ventricular arrhythmia (contraction frequency for ventricles – 75–133 beats per minute). Repolarization
disturbances in the lateral wall of the left ventricle probably as an
indication of coronary insufficiency on the background of hypertrophy of the left ventricle.

Date of entry – 28.03.2005
Date of birth – 08.07.1926
Full name – XX10 28-03-2005 11-05-00

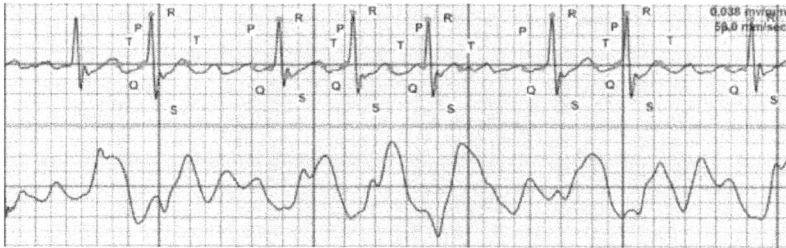

Diagnosis resulted from phase analysis.
The volume-related parameters of function of the heart and the associated blood vessels are within normal values.

Doctor's diagnosis.
The function of the aortic valve – norm.

Anatomic peculiarities of aortic valve in the tension phase indicated on the ECG curve – norm.

Anatomic peculiarities of the aortic valve in the full opening phase indicated on the ECG curve – norm.

Flabbiness of aorta – apparent. The pressure increases in the slow ejection phase step by step and, upon reaching its maximum, decreases up to its minimum.

Condition of elasticity of aorta – S. "Flabbiness of aorta" above.

Condition of contraction function of ventricular septum – norm.

Condition of contraction function of myocardium – the function is diminished. The amplitude of the lower portion of the QRS complex is a half of the norm.

Condition of venous blood flow – norm.

Condition of function of the lungs – norm.

Potential stroke problem – no sudden hemopulses are available. Potential stroke problem is not available.

ORTHOSTATIC TEST WAS NOT PERFORMED.

SV (ml) – Stroke volume
SV(ml)=55.29 %=0.00   42.29   96.42

MV (l) – Minute volume
MV(l)=5.40 %=0.00   4.13   9.41

PV1 (ml) – Volume of blood in phase of rapid and slow filling of left (aortic) ventricle of heart
PV1(ml)=28.13 %=0.00   19.28   43.12

PV2 (ml) – Volume of blood flowing into left (aortic) ventricle of heart during systole
PV2(ml)=27.16 %=0.00   23.01   53.30

PV3 (ml) – Volume of blood ejected by left (aortic) ventricle of heart during rapid ejection phase
PV3(ml)=32.82 %=0.00   25.08   57.30

PV4 (ml) – Volume of blood ejected by left (aortic) ventricle of heart during slow ejection phase
PV4(ml)=22.48 %=0.00   17.21   39.12

PV5 (ml) – Volume of blood transferred by ascending aorta functioning as peristaltic pump
PV5(ml)=7.67 %=0.00   6.67   10.91

	QRS	RS	QT	PQ	TT	SV (ml)	MV (1)	PV1 (ml)	PV2 (ml)	PV3 (ml)	PV4 (ml)	PV5 (ml)	HR
Min	0.080	0.035	0.314	0.066	0.615	42.3	4.1	19.3	23.0	25.1	17.2	6.7	97.6
Max	0.110	0.060	0.314	0.066	0.615	96.4	9.4	43.1	53.3	57.3	39.1	10.9	97.6
Average	0.087	0.042	0.299	0.061	0.615	55.3	5.4	28.1	27.2	32.8	22.5	7.7	97.6
0	0.084	0.038	0.270	0.025	0.434	44.6	6.2	-4.9	49.5	26.5	18.1	6.1	138.2
1	0.097	0.042	0.341	0.114	0.889	58.5	3.9	34.9	23.6	34.7	23.8	8.7	67.5
2	0.084	0.042	0.282	0.025	0.434	54.3	7.5	-27.5	81.7	32.2	22.1	7.3	138.2
3	0.072	0.042	0.240	0.013	0.455	51.0	6.7	29.2	21.8	30.3	20.7	6.3	131.8
4	0.110	0.042	0.346	0.110	0.873	57.8	4.0	33.4	24.4	34.3	23.5	8.4	68.8
5	0.084	0.042	0.257	0.025	0.422	51.5	7.3	-5.9	57.4	30.6	20.9	6.4	142.3
6	0.084	0.042	0.333	0.114	0.881	58.9	4.0	35.9	22.9	34.9	23.9	8.8	68.1
7	0.080	0.038	0.266	0.021	0.438	44.6	6.1	4.8	39.8	26.5	18.1	6.1	136.9
8	0.084	0.042	0.253	0.021	0.443	51.0	6.9	14.8	36.1	30.3	20.7	6.3	135.6
9	0.072	0.042	0.211	0.021	0.443	47.0	6.4	26.3	20.7	27.9	19.1	5.2	135.6
10	0.110	0.042	0.354	0.101	0.873	58.5	4.0	34.4	24.2	34.7	23.8	8.7	68.8
11	0.084	0.042	0.261	0.013	0.430	52.0	7.3	9.5	42.4	30.9	21.1	6.6	139.6
12	0.089	0.042	0.291	0.152	0.864	54.7	3.8	30.5	24.2	32.5	22.2	7.4	69.4
13	0.080	0.038	0.270	0.017	0.443	45.0	6.1	7.8	37.2	26.7	18.3	6.2	135.6
14	0.118	0.042	0.358	0.097	0.885	58.2	3.9	34.4	23.8	34.5	23.7	8.5	67.8
15	0.093	0.042	0.274	0.017	0.430	52.5	7.3	-6.1	58.6	31.1	21.3	6.7	139.6
16	0.067	0.042	0.223	0.017	0.447	49.4	6.6	28.9	20.4	29.3	20.0	5.9	134.3
17	0.076	0.042	0.312	0.025	0.548	57.8	6.3	30.4	27.4	34.3	23.5	8.4	109.5
18	0.089	0.042	0.350	0.093	0.759	59.9	4.7	32.2	27.6	35.5	24.4	9.1	79.1
19	0.089	0.042	0.396	0.051	0.662	63.1	5.7	27.2	35.9	37.4	25.7	10.4	90.7
20	0.093	0.042	0.354	0.084	0.658	59.9	5.5	22.5	37.4	35.5	24.4	9.1	91.2
21	0.084	0.046	0.257	0.055	0.451	59.9	8.0	-5.0	64.9	35.6	24.3	7.1	133.0
22	0.089	0.042	0.278	0.160	0.851	53.4	3.8	29.0	24.4	31.7	21.7	7.0	70.5
23	0.080	0.042	0.261	0.021	0.464	52.5	6.8	20.7	31.8	31.1	21.3	6.7	129.4
24	0.089	0.046	0.278	0.164	0.856	62.2	4.4	33.8	28.4	37.0	25.3	7.8	70.1
25	0.080	0.042	0.451	0.021	0.653	66.7	6.1	26.3	40.4	39.6	27.2	11.8	91.8

Comparative analysis of example No. 10 from practice

Phase duration.

The QRS complexes are equal in their duration.

A difference in the phase duration between the QT phases is 0.061s.

Diagnoses.

Classical diagnosis option: Atrial fibrillation with ventricular arrhythmia. Repolarization disturbances in the lateral wall of the left ventricle.
The phase analysis shows the following.
There are two factors of great importance which should require a special attention. The first factor is arrhythmia. It should be analyzed in combination with another factor – the flabbiness of the aorta. This may be useful in order to determine the effect of the arrhythmia on hemodynamics as a whole. When analyzing the arterial pressure in the heart cycle, it should be noted that in the long cycle, when a sharp decrease in the arterial pressure due to the aorta flabbiness is available, a heavy hindrance to the venous flow is detected, too. This results in a significant increase in the arterial pressure in this phase. The short cycles do not show such phenomena. In the short cycles, the arterial pressure waves are fully within the norms. It seems that extrasystole occurs as a protective function that is capable of providing the required blood flow within the venous circulation system actually due to the atrial functioning. It must occur because the complete cycles described above are practically not capable of pumping blood in the venous circulation system.
Therefore, a conclusion could be made that the therapy target should be the normalization of the functioning of the venous system and the minimization of the aorta flabbiness. The function of the heart is within the relevant norms in all function phases.

Example No. 11 from Practice

Antecedent Anamnesis
Patient (full name: XX10), male, born 1939. Periodically registered complaints about hypertension; suffered from exudative pleurisy and bronchopneumonia.

Diagnosis
Coronary heart disease (CHD), atherosclerotic cardiosclerosis.

Clinical Examination
Complaints: weakness at physical activity. Basic data: weight – 69 kgs, stature – 160 cm; body mass index (BMI) – 26.75 kg/m2 (e x c e s -sive body mass).

Dynamic data: arterial pressure (AP) 150/90, 120/80, 110/70, 120/80. Heart frequency: 76 beats per minute, breathing frequency: 16–18 breath movements per minute.

Auscultation: heart tones are damped down. Breathing in lungs: vesicular. Abdomen is soft, no pains available. The liver borders are within the norm.

Tests
Laboratory Testing.

General blood test: leukocytes – $7.1*10*9$ (leukocyte formula shows no pathology), erythrocytes – $4.52*10*12$, hemoglobin 128 g/l, cyclic immunocomplex – 1, eryth-rocyte sedimentation rate (ESR) – 5 mm/h, prothrombin – 106%, AST – 0.25 μmole/l.h, ALT – 0.35 μmole/l.h, glucose – 6.7 μmole/l, C-reactive protein – negative. Urine analysis: no pathology.

329

Instrumentation testing.
ECG: P – 0.08; PQ – 0.14; QRS – 0.08;
QRST – 0.40; R–R – 0.95
Heart frequency – 63 beats per minute. There are diffuse changes in myocardium on the background of sinus rhythm available.

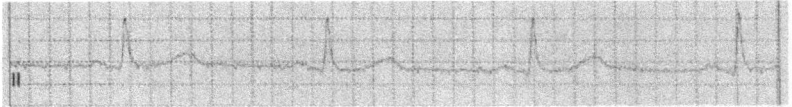

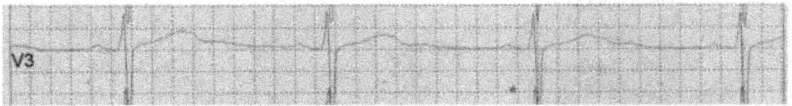

Date of entry – 24.03.2005
Date of birth – 09.10.1939
Full name – XX11 24-03-2005 13-41-25

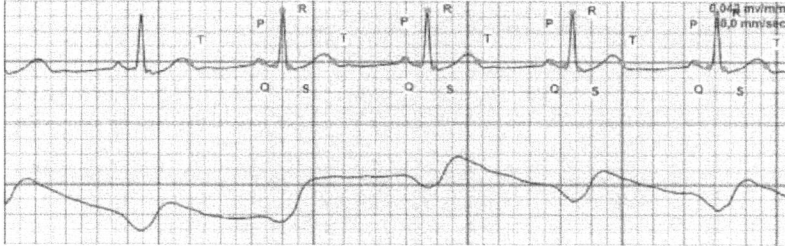

Diagnosis resulted from phase analysis.
The volume-related parameters of function of the heart and the associated blood vessels are within normal values.

Doctor's diagnosis.
The function of the aortic valve – significantly diminished.
There is a contraction of the ventricular septum already in the phase QR available, next, there is a contraction of myocardium in the phase RS available, the valve is leaking blood into the aorta, and that is the cause of the minimum arterial pressure available there.
Anatomic peculiarities of aortic valve in the tension phase indicated on the ECG curve. A changed condition due to the fact that there is no wave S available.
Anatomic peculiarities of the aortic valve in the full opening phase indicated on the ECG curve – norm.

Flabbiness of aorta – not available.

Condition of elasticity of aorta – norm.
Condition of contraction function of ventricular septum – norm.

Condition of contraction function of myocardium – the function is considerably diminished, and that is an indication of the maximum passivity of the myocardium contraction.

Condition of venous blood flow – slightly hindered.

Condition of function of the lungs – norm.

Potential stroke problem – no sudden hemopulses are available. Potential stroke problem is not available.

ORTHOSTATIC TEST WAS NOT PERFORMED.

SV (ml) – Stroke volume

| SV(ml)=65.23 %=0.00 | 44.51 | | 104.17 |

MV (l) – Minute volume

| MV(l)=4.25 %=0.00 | 2.90 | | 6.78 |

PV1 (ml) – Volume of blood in phase of rapid and slow filling of left (aortic) ventricle of heart

| PV1 (ml)=42.55 %=0.00 | 29.79 | | 69.05 |

PV2 (ml) – Volume of blood flowing into left (aortic) ventricle of heart during systole

| PV2 (ml)=22.68 %=0.00 | 14.72 | | 35.12 |

PV3 (ml) – Volume of blood ejected by left (aortic) ventricle of heart during rapid ejection phase

| PV3 (ml)=38.69 %=0.00 | 26.39 | | 61.86 |

PV4 (ml) – Volume of blood ejected by left (aortic) ventricle of heart during slow ejection phase

| PV4 (ml)=26.54 %=0.00 | 18.12 | | 42.31 |

PV5 (ml) – Volume of blood transferred by ascending aorta functioning as peristaltic pump

| PV5 (ml)=10.15 %=0.00 | 7.54 | | 13.00 |

	QRS	RS	QT	PQ	TT	SV (ml)	MV (1)	PV1 (ml)	PV2 (ml)	PV3 (ml)	PV4 (ml)	PV5 (ml)
Min	0.080	0.035	0.355	0.085	0.922	44.5	2.9	29.8	14.7	26.4	18.1	7.5
Max	0.110	0.060	0.355	0.085	0.922	104.2	6.8	69.0	35.1	61.9	42.3	13.0
Average	0.091	0.044	0.373	0.088	0.922	65.2	4.2	42.5	22.7	38.7	26.5	10.1
0	0.093	0.046	0.375	0.080	0.927	71.9	4.7	48.2	23.7	42.6	29.2	10.9
1	0.093	0.042	0.375	0.084	0.935	61.3	3.9	40.5	20.8	36.4	24.9	9.7
2	0.088	0.042	0.371	0.088	0.939	61.3	3.9	40.6	20.7	36.4	24.9	9.7
3	0.093	0.046	0.371	0.088	0.935	71.5	4.6	47.2	24.3	42.4	29.1	10.8
4	0.093	0.046	0.371	0.088	0.927	71.5	4.6	46.9	24.6	42.4	29.1	10.8
5	0.088	0.042	0.375	0.088	0.931	61.6	4.0	40.4	21.2	36.5	25.1	9.8
6	0.088	0.042	0.371	0.088	0.910	61.3	4.0	39.7	21.6	36.4	24.9	9.7
7	0.093	0.042	0.379	0.088	0.931	61.6	4.0	39.9	21.7	36.5	25.1	9.8
8	0.088	0.042	0.375	0.088	0.922	61.6	4.0	40.2	21.4	36.5	25.1	9.8
9	0.088	0.042	0.375	0.093	0.918	61.6	4.0	39.5	22.1	36.5	25.1	9.8
10	0.093	0.046	0.371	0.088	0.918	71.5	4.7	46.6	24.9	42.4	29.1	10.8
11	0.093	0.046	0.375	0.088	0.939	71.9	4.6	47.4	24.4	42.6	29.2	10.9
12	0.088	0.042	0.375	0.088	0.927	61.6	4.0	40.3	21.3	36.5	25.1	9.8
13	0.088	0.042	0.375	0.088	0.897	61.6	4.1	39.3	22.3	36.5	25.1	9.8
14	0.093	0.042	0.371	0.088	0.889	61.0	4.1	38.5	22.5	36.2	24.8	9.6
15	0.093	0.046	0.371	0.088	0.905	71.5	4.7	46.2	25.4	42.4	29.1	10.8

Comparative analysis of example No. 11 from practice

Phase duration.

The QRS complexes are equal in their duration.

A difference in the phase duration between the PQ phases is 0.052s.

A difference in the phase duration between the QT phases is 0.027s.

Diagnoses.

Classical diagnosis option: diffuse changes in myocardium.

The phase analysis shows the following.

It is evident that there is no wave S is available. The hemodynamic parameters are within the norm being slightly under the normal values. There is a minor hindrance in venous flow detected, but, actually, the venous circulation is within the norm. The patient suffers from complaints about an increased arterial pressure (AP). That may be caused by a weak functioning of the valve which is leaking blood from the phase of the ventricular contraction. According to the RHEO, the AP may reach 95 mm Hg and be higher.

The therapy target should be normalization of the aortic valve function.

Example No. 12 from Practice

Antecedent Anamnesis
Patient (full name: XX12), male, born 1953. Since 1974 registered as tuberculous patient. 1976 operated upon tuberculoma of light lung. 1991 tuberculosis recurrence registered. 1995 the patient came through myocardial infarction.

Diagnosis
Disseminated pulmonary tuberculosis, infiltrative attack & destruction phase (chronicity), IIA, MBT+, state after seg-mental resection of the upper lobe of the right lung. Coronary heart disease (CHD). Exertional & rest angina pectoris. Functional Class, risc II, No.

Clinical Examination
Complaints: dyspnea at physical activity, cough accompanied by minor sputum discharge, pains in the heart area and weakness.
Basic data: weight – 57 kgs, stature – 175 cm; body mass index (BMI) – 18.41 kg/m2 (body mass deficit).
Dynamic data: arterial pressure (AP) 110/ 8 0 , 12 0/8 0.
Heart frequency: 72 beats per minute, breathing frequency: 20-22 breath movements per minute.
Auscultation: heart tones are damped down; extrasystole.
Breathing in lungs: rough, single dry rales. Abdomen: soft, pains in the right hypo-chondrium are available. The liver borders extend 3 cm beyond the edge of the costal margin.

Tests

Laboratory Testing.

General blood test: leukocytes – 10*10*9, erythrocytes – 4.52*10*12, hemoglobin 140 g/l, erythrocyte sedimentation rate (ESR) – 23 mm/h, prothrombin – 106%, glucose – 5.0, bilirubin, total – 8.6, bilirubin conjugated – 0.0; AST – 0.5 μmole/l.h, ALT – 0.5 μmole/l.h, bacterial inoculation – rich growth of tuberculosis mycobacteria. Urine analysis: no pathology.

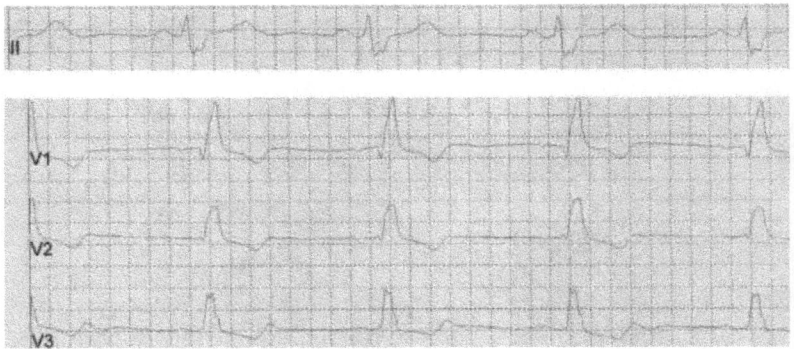

Instrumentation testing.

ECG: P-0.09; PQ – 0.12; QRS – 0.14; QRST – 0.43; R–R – 0.94 Heart frequency – 64 beats per minute. Sinus rhythm is available. Myocardial hypertrophy of the right atrium. Complete right bundle-branch block. Apparent changes in myocardium.

Radiographic examination:

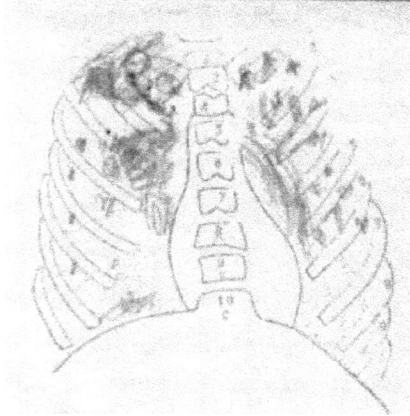

There are post-operation sutures in the upper lobe of the right lung detected; a cirrhosis area with cavities in the lungs is available; rough pneumosclerosis. Clustered random-sized nidi and small-sized caseational foci with a diameter up to 1.5 cm are detected. There are random-sized nidi and small tuberculomas with clear contours available in the upper lobe on the left side. The roots are fibrously deformed, the borders of the heart are within the norm.

Date of entry – 14.05.2005
Date of birth – 01.05.1953
Full name – XX12 14-05-05 09-10-38 am

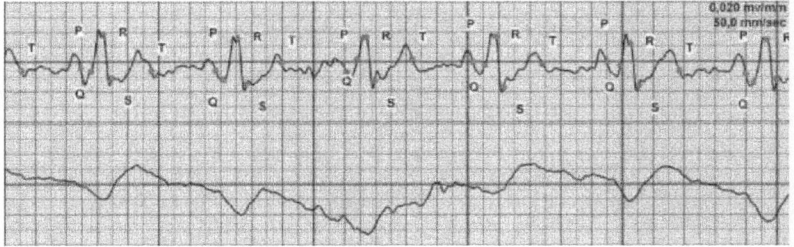

Diagnosis resulted from phase analysis.
The volume-related parameters of function of the heart and the associated blood vessels are within normal values.

Doctor's diagnosis.
The function of the aortic valve – diminished. There is a contraction of myocardium in the phase RS up to the wave point S available, and the valve is leaking blood into the aorta. This is the cause of an increase in the minimum arterial pressure (AP).
Anatomic peculiarities of aortic valve in the tension phase indicated on the ECG curve. The wave S on the ECG curve has a small amplitude. The blood enters the aorta in this phase. This is a pathology.
Anatomic peculiarities of the aortic valve in the full opening phase indicated on the ECG curve. A changed condition indicated as smoothed wave Lj is available. But the pressure increasing rate, according to the RHEO, is small. By the valve opening, the pressure in the aorta reaches 50% of the maximum one.

Flabbiness of aorta –not available.

Condition of elasticity of aorta – slightly reduced.

338

The apex on the RHEO is slightly "beveled" (slightly flattened).

Condition of contraction function of ventricular septum – there is a pathology of the ventricular septum available, and its indication is a bifurcation of the wave R on the ECG curve.

Condition of contraction function of myocardium – the function is significantly diminished. The amplitude of the lower portion of the QRS complex is a minimum, and that indicates the maximum passivity of the myocardium contraction.

Condition of venous blood flow – the venous flow is hindered. According to the RHEO, after the dicrotic valley, the curve portion is horizontal.

Condition of function of the lungs – the function of the lungs is diminished. A minor modulation of the RHEO curve by breathing is available.

Potential stroke problem – no sudden hemopulses are available. Potential stroke problem is not available.

ORTHOSTATIC TEST WAS NOT PERFORMED.

Additional information. Special attention should be paid to the "dip" of the wave PQ on the ECG curve. It is known from praxis, that the shape of such a phase could be considered as a critical symptom. As a rule, this phenomenon is accompanied by an increase in the amplitude of the wave P. All these facts are available in this diagnostics case. This is of great importance because this signal shape in this phase is developing in such an anatomical segment of the heart to which the main heart structures are connected: the valves, the ventricular septum and the atrii. The "dip" on the curve is an indication of both functional weakness of the heart as a whole and heart overloading, and this statement is supported by the phase analysis.

SV (ml) – Stroke volume

| SV(ml)=92.12 X=0.00 | 43.80 | | 101.68 |

MV (l) – Minute volume

| MV(l)=6.51 X=0.00 | 3.09 | | 7.18 |

PV1 (ml) – Volume of blood in phase of rapid and slow filling of left (aortic) ventricle of heart

| PV1(ml)=57.05 X=0.00 | 28.31 | | 65.01 |

PV2 (ml) – Volume of blood flowing into left (aortic) ventricle of heart during systole

| PV2(ml)=35.07 X=0.00 | 15.48 | | 36.67 |

PV3 (ml) – Volume of blood ejected by left (aortic) ventricle of heart during rapid ejection phase

| PV3(ml)=54.69 X=0.00 | 25.97 | | 60.39 |

PV4 (ml) – Volume of blood ejected by left (aortic) ventricle of heart during slow ejection phase

| PV4(ml)=37.42 X=0.00 | 17.83 | | 41.28 |

PV5 (ml) – Volume of blood transferred by ascending
aorta functioning as peristaltic pump

| PV5(ml)=11.96 X=0.00 | 7.25 | | 12.30 |

	QRS	RS	QT	PQ	TT	SV (ml)	MV (1)	PV1 (ml)	PV2 (ml)	PV3 (ml)	PV4 (ml)	PV5 (ml)	HR
Min	0.080	0.035	0.341	0.082	0.849	43.8	3.1	28.3	15.5	26.0	17.8	7.3	70.6
Max	0.110	0.060	0.341	0.082	0.849	101.7	7.2	65.0	36.7	60.4	41.3	12.3	70.6
Average	0.130	0.056	0.375	0.066	0.849	92.1	6.5	57.0	35.1	54.7	37.4	12.0	70.6
0	0.138	0.055	0.394	0.063	0.839	90.8	6.5	53.8	36.9	53.9	36.9	12.1	71.5
1	0.147	0.055	0.377	0.050	0.843	87.2	6.2	54.4	32.8	51.8	35.4	11.1	71.2
2	0.109	0.055	0.357	0.071	0.843	89.6	6.4	57.5	32.1	53.2	36.4	11.8	71.2
3	0.122	0.050	0.373	0.075	0.839	79.2	5.7	47.2	31.9	47.0	32.2	10.9	71.5
4	0.117	0.055	0.382	0.088	0.885	91.9	6.2	55.7	36.1	54.5	37.3	12.5	67.8
5	0.151	0.063	0.386	0.050	0.864	110.4	7.7	70.3	40.1	65.6	44.8	13.1	69.4
6	0.134	0.055	0.390	0.059	0.856	90.8	6.4	56.4	34.3	53.9	36.9	12.1	70.1
7	0.122	0.059	0.344	0.067	0.826	96.7	7.0	61.5	35.2	57.4	39.3	11.6	72.6

341

Comparative analysis of example No. 12 from practice

Phase duration.

The QRS complexes are equal in their duration.

A difference in the phase duration between the PQ phases is 0.056s.

A difference in the phase duration between the QT phases is 0.086s.

Diagnoses.

Classical diagnosis option: Myocardial hypertrophy of the right atrium. Complete right bundle-branch block.

The phase analysis shows the following.

Please, note the text of the "additional information" given above in the Clause "Doctor's diagnosis". Please, note other statements in this Clause, too. The complete right bundle-branch block should be additionally commented, too. As the apex of the wave R should be considered the second maximum registered on the ECG curve. The fact that the first apex is located close to the second one indicates that a conductivity pathology is located close to the heart apex. The "dip" in the PQ phase is very dangerous. But irrespective of these facts, all hemodynamic parameters are relatively within the norm.
The therapy should be a comprehensive therapy course, and the therapy target should be a step-by-step removal of the above problems; first of all, it should be aimed at strengthening of the anatomical basement of the heart.

Example No. 13 from Practice

Antecedent Anamnesis
Patient (full name: XX13), male, born 1954. Since 2004 registered as tuberculous patient. 2003 the patient came through myocardial infarction.

Diagnosis
Pulmonary fibrous cavernous tuberculosis of upper lobes of both lungs, infiltration & seeding phase, IIA, MBT+. Coronary heart disease (CHD). Exer-tional & rest angina pectoris. Functional Class III. Post-infarction cardiosclerosis. Chronic cardiac insufficiency, degree II.

Clinical Examination
Complaints: dyspnea at physical activity, cough accompanied by sputum, pains in the heart area and weakness, low grade fever.
Basic data: weight – 65 kgs, stature – 175 cm; body mass index (BMI) – 21.02 kg/m2 (body mass norm).
Dynamic data: arterial pressure (AP) 14 0/ 8 0 , 16 0/10 0 .
Heart frequency: 79 beats per minute, breathing frequency: 18–20 breath movements per minute.
Auscultation: heart tones are damped down. Breathing in lungs: rough, single moist and dry rales.
Abdomen: soft, pains in the right hypo-chondrium are available. The liver borders are at the edge of the costal margin.

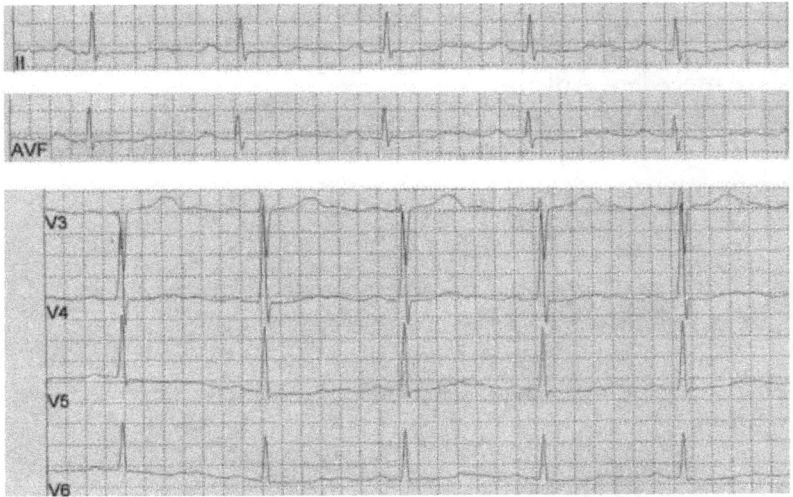

Tests

Laboratory Testing.

General blood test: leukocytes – 6.5*10*9, erythrocytes – 4.52*10*12, hemoglobin 140 g/l, erythrocyte sedimentation rate (ESR) – 30 mm/h, prothrombin – 106%, glucose – 3.1, bilirubin, total – 6.2, bilirubin conjugated – 0.0; AST – 0.42 μmole/l.h, ALT – 0.17 μmole/l.h, bacterial inoculation – rich growth of tuberculosis mycobacteria. Urine analysis: no pathology.

Instrumentation testing. ECG: P–0.10; PQ – 0.16; QRS – 0.08; QRST – 0.38; R–R – 0.71 Heart frequency – 85 beats per minute. Sinus rhythm is available. Myocardial hypertrophy of the right atrium. Repo-larization disturbances in the lower and anterolateral wall of the left ventricle are available, latent subendocardial insufficiency is possibly available.

Radiographic examination:

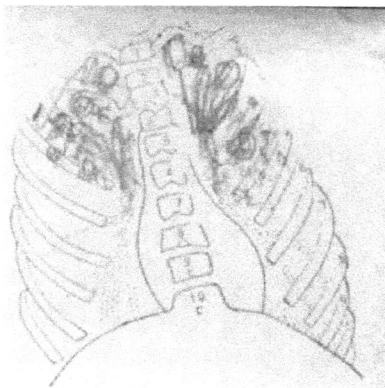

In the upper lobes in both lungs, against the cirrhosis background, multiple nidi and small foci without clear contours are available; multiple destruction cavities with a diameter from 1.0 to 3.0 cm are available; the roots are pulled and deformed.

Date of entry – 03.06.2005
Date of birth – 13.10.1954
Full name – XX13 03-06-05 09-47-54 am

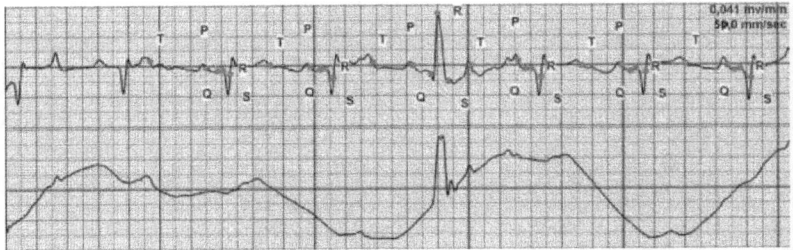

Diagnosis resulted from phase analysis.
SV (ml.l) – Within the norm.
MV (l) – Within the norm.
PV1 (l) – Within the norm.
PV2 (l) – Within the norm.
PV3 (l) – Within the norm.
PV4 (l) – Within the norm.
PV5 (l) – There is an increase in the blood volume pumped by the ascending aorta as peristaltic pump. The tonus of the ascending aorta is strengthened. The consequence is a decrease in the load on the left ventricle due to a reduced resistance of the aorta. The volume-related parameters of the function of the heart and the associated blood vessels are within the normal values.

Doctor's diagnosis.
The function of the aortic valve – norm.

Anatomic peculiarities of aortic valve in the tension phase indicated on the ECG curve. A changed condition as an increase in the amplitude of the wave SL is available.

346

Anatomic peculiarities of the aortic valve in the full opening phase indicated on the ECG curve. A changed condition as the smoothed wave Lj is available. The pressure in the aorta does not reach practically its normal value. This is an indication of the weakness of the ventricles.

Flabbiness of aorta – apparent. There is not an increase, but a sharp decrease in the pressure in the slow ejection phase available.

Condition of elasticity of aorta – slightly reduced. The apex of the rheogram is flat.

Condition of contraction function of ventricular septum – the function is significantly diminished.

The apex R amplitude is less than the T wave amplitude.

Condition of contraction function of myocardium – the function is diminished. The amplitude of the lower portion of the QRS complex is a half of the norm.

Condition of venous blood flow – norm.

Condition of function of the lungs – the function of the lungs is significantly diminished. A heavy modulation of the rheogram curve by breathing is available.
Potential stroke problem – hemopulses with flat apeci are available. Susceptibility to a disturbance in the blood flow structure and effect of erythrocytes fusion leading to thrombosis and then to ruptures of cerebral blood vessels.

ORTHOSTATIC TEST WAS NOT PERFORMED.

SV (ml) – Stroke volume

| SV(ml)=102.85 %=8.14 | 41.92 | | 95.11 |

MV (l) – Minute volume

| MV(l)=8.95 %=8.14 | 3.65 | | 8.27 |

PV1 (ml) – Volume of blood in phase of rapid and slow filling of left (aortic) ventricle of heart

| PV1(ml)=58.22 %=9.32 | 23.81 | | 53.26 |

PV2 (ml) – Volume of blood flowing into left (aortic) ventricle of heart during systole

| PV2(ml)=44.63 %=6.63 | 18.11 | | 41.85 |

PV3 (ml) – Volume of blood ejected by left (aortic) ventricle of heart during rapid ejection phase

| PV3(ml)=61.08 %=8.05 | 24.86 | | 56.52 |

PV4 (ml) – Volume of blood ejected by left (aortic) ventricle of heart during slow ejection phase

| PV4(ml)=41.77 %=8.26 | 17.06 | | 38.59 |

PV5 (ml) – Volume of blood transferred by ascending
aorta functioning as peristaltic pump

| PV5(ml)=12.97 %=22.64 | 6.54 | | 10.58 |

	QRS	RS	QT	PQ	TT	SV (ml)	MV (1)	PV1 (ml)	PV2 (ml)	PV3 (ml)	PV4 (ml)	PV5 (ml)
Min	0,080	0,035	0,307	0,072	0,690	41,9	3,6	23,8	18,1	24,9	17,1	6,5
Max	0,110	0,060	0,307	0,072	0,690	95,1	8,3	53,3	41,9	56,5	38,6	10,6
Average	0,093	0,059	0,341	0,066	0,690	102,8	8,9	58,2	44,6	61,1	41,8	13,0
0	0,080	0,046	0,341	0,072	0,779	70,0	5,4	44,9	25,1	41,5	28,5	10,2
1	0,084	0,055	0,337	0,067	0,661	90,9	8,3	48,0	42,9	54,0	37,0	12,1
2	0,109	0,067	0,253	0,118	0,636	97,3	9,2	42,8	54,5	57,9	39,4	8,4
3	0,105	0,063	0,366	0,046	0,720	116,0	9,7	72,2	43,8	68,9	47,1	14,6
4	0,097	0,063	0,358	0,059	0,678	116,0	10,3	63,3	52,7	68,9	47,1	14,6
5	0,105	0,072	0,362	0,046	0,678	140,6	12,4	83,1	57,5	83,5	57,1	16,4
6	0,080	0,059	0,341	0,076	0,691	103,8	9,0	57,7	46,1	61,6	42,2	13,5
7	0,084	0,055	0,350	0,063	0,691	92,6	8,0	53,0	39,6	55,0	37,6	12,6
8	0,076	0,051	0,333	0,067	0,674	80,3	7,2	45,4	34,9	47,7	32,6	11,2
9	0,109	0,063	0,366	0,042	0,691	115,2	10,0	68,1	47,1	68,4	46,8	14,3

Comparative analysis of example No. 13 from practice

Phase duration.

A difference in the duration of the QRS complexes is 0.029 s.

A difference in the phase duration between the PQ phases is 0.118s.

A difference in the phase duration between the QT phases is 0.014s.

Diagnoses.

Classical diagnosis option: Myocardial hypertrophy of the right atrium. Repolarization disturbances in the lower and anterolateral wall of the left ventricle are available.

The phase analysis shows the following.

The main problem of this patient is the fact that there is no wave R on the ECG curve available. The ventricular septum does not function at all. There is a heavy modulation of the RHEO by breathing waves detected.

The cause is the weakness in lung functioning. The hemopulses of rectangular shape are registered. This is an indication of potential stroke development. The phase analysis generated automatically shows a great load on the aorta.

When the arterial pressure hemopulses are detected, retroversion or anteversion are not recommended for this patient; the patient should avoid pain development in the occiput. The R wave "dip" is an indication of swelling of the cell membranes. Depending on the position of the body, it may be developed or not. Therefore, an orthostatic test should be recommended in this case. For the above

patient this test was not carried out. An additional difficulty should
be weakness of the lungs.

Example No. 14 from Practice

Antecedent Anamnesis
Patient (full name: XX14), male, born 1974. Since 2005 registered as tuberculous patient. Registered congenital malformation of heart – defect of ventricular septum.

Diagnosis
Disseminated pulmonary tuberculosis of lungs, destruction phase complicated by right-side exudative pleurisy, IA, TMB+.

Clinical Examination
Complaints: dyspnea at physical activity, dry cough, weakness, hyperhidrosis, febrile temperatures.
Basic data: weight – 66 kgs, stature – 178 cm; body mass index (BMI) – 20.63 kg/m2 (body mass norm).

Dynamic data: arterial pressure (AP) 110/60, 120/80.
Heart frequency: 68 beats per minute, breathing frequency: 19–21 breath movements per minute.
Auscultation: heart tones are damped down.
Breathing in lungs: weakened on the right side below the scapula angle, dry rales on both sides.
Abdomen: soft, no pains available. The liver borders are at the edge of the costal margin.

Tests
Laboratory Testing.
General blood test: leukocytes – 4.0*10*9, erythrocytes – 4.52*10*12, hemoglobin 140 g/l, erythrocyte sedimentation rate (ESR) – 56 mm/h, prothrombin – 106%, glucose – 3.0; bilirubin, total – 10.0, bilirubin conjugated – 0.0; AST – 0.42 μmole/l.h, ALT – 0.5 μmole/l.h, bacterial inoculation – scare growth of tuberculosis

mycobacteria. Urine analysis: protein – 0.099, individual erythro-cytes, leukocytes 1–3, individual granular cylinders.

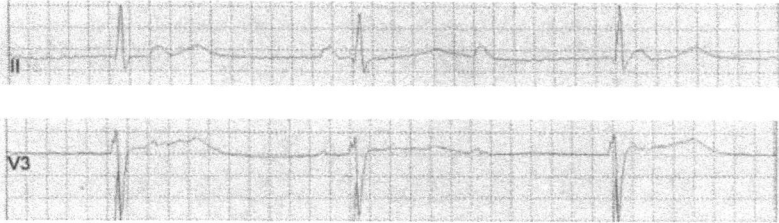

Instrumentation testing. ECG: P – 0.08; PQ – 0.37; QRS -0.09, QRST – 0.46; R–R – 1.20. Heart frequency – 50 beats per min-ute. Moderate sinus bradycardia. Myocardial hypertrophy of right atrium. Complete third degree block (proximal). An increase in load on the right atrium. Signs of myo-cardial hypertrophy of the left ventricle. Repolarization disturbances in myocardium probably as manifestation of coronary insufficiency. Dismetabolic changes in myocardium.

Radiographic examination:

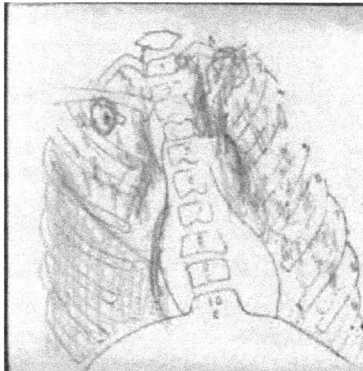

Right side: in the upper lobe fibrosis, polymorphous nidi in S2, a cavity with a diameter of 2.0 cm, below rib IV – an intensive homogenous shadow with the top inclined border line. Left side: almost throughout the total length, against the background of an enhanced lung pattern, – polymorphous nidi, in S1, S2 – a deformed cavity of 4.0–2.0 cm. The roots are fibrous.

Date of entry – 18.04.2005
Date of birth – 14.08.1974
Full name – XX14 18-04-05 12-58-25

Diagnosis resulted from phase analysis.
The volume-related parameters of function of the heart and the associated blood vessels are within normal values.

Doctor's diagnosis.
The function of the aortic valve – norm.

Anatomic peculiarities of aortic valve in the tension phase indicated on the ECG curve. A changed condition as an increase in the SL wave amplitude is available.

Anatomic peculiarities of the aortic valve in the full opening phase indicated on the ECG curve. A changed condition as smoothed wave Lj is available.

Flabbiness of aorta – not available.

Condition of elasticity of aorta – norm. The apex on the rheogram curve is parabolic and is positioned left referred to the wave point T on the ECG curve.
Condition of contraction function of ventricular septum – considerably diminished. The R apex amplitude is at the same level as the amplitude of the T wave is.

Condition of contraction function of myocardium – norm.

Condition of venous blood flow – hindered.

Condition of function of the lungs – diminished.

Potential stroke problem – no sudden hemopulses are available. Potential stroke problem is not available.

ORTHOSTATIC TEST WAS NOT PERFORMED.

Features detected. The pulse is very slow. There is practically no wave T available on the ECG curve. The "floating" wave P is registered. This occurs very seldom. The shift of this wave is connected with the slow pulse, and this wave makes it possible to increase the pressure in the aorta up to mean values. But it is not the wave U since the latter changes its position from cycle to cycle.

SV (ml) – Stroke volume

| SV(ml)=65.98 %=0.00 | 45.76 | | 108.53 |

MV (l) – Minute volume

| MV(l)=3.71 %=0.00 | 2.57 | | 6.10 |

PV1 (ml) – Volume of blood in phase of rapid and slow filling of left (aortic) ventricle of heart

| PV1(ml)=36.74 %=0.00 | 32.22 | | 75.77 |

PV2 (ml) – Volume of blood flowing into left (aortic) ventricle of heart during systole

| PV2(ml)=29.24 %=0.00 | 13.54 | | 32.76 |

PV3 (ml) – Volume of blood ejected by left (aortic) ventricle of heart during rapid ejection phase

| PV3(ml)=39.17 %=0.00 | 27.13 | | 84.43 |

PV4 (ml) – Volume of blood ejected by left (aortic) ventricle of heart during slow ejection phase

| PV4(ml)=26.82 %=0.00 | 18.63 | | 44.10 |

PV5 (ml) – Volume of blood transferred by ascending
aorta functioning as peristaltic pump

| PV5(ml)=8.91 %=0.00 | 8.06 | | 14.23 |

356

	QRS	RS	QT	PQ	TT	SV (ml)	MV (1)	PV1 (ml)	PV2 (ml)	PV3 (ml)	PV4 (ml)	PV5 (ml)
Min	0.080	0.035	0.382	0.090	1.068	45.8	2.6	32.2	13.5	27.1	18.6	8.1
Max	0.110	0.060	0.382	0.090	1.068	108.5	6.1	75.8	32.8	64.4	44.1	14.3
Average	0.087	0.046	0.308	0.229	1.068	66.0	3.7	36.7	29.2	39.2	26.8	8.9
0	0.097	0.046	0.358	0.328	1.120	70.0	3.7	31.3	38.7	41.5	28.5	10.2
1	0.063	0.046	0.270	0.013	0.985	64.2	3.9	58.7	5.6	38.1	26.1	8.4
2	0.097	0.046	0.337	0.354	1.103	68.0	3.7	28.7	39.2	40.3	27.6	9.5
3	0.088	0.046	0.295	0.029	1.049	64.2	3.7	53.2	11.0	38.1	26.1	8.4
4	0.088	0.046	0.282	0.421	1.082	62.7	3.5	23.7	39.0	37.2	25.5	7.9

Comparative analysis of example No. 14 from practice

Phase duration.

The QRS complexes are equal in their duration.

A difference in the phase duration between the PQ phases is 0.141s.

A difference in the phase duration between the QT phases is 0.158s.

Diagnoses.

Classical diagnosis option: Myocardial hypertrophy of the right atrium. Complete third degree block.

The phase analysis shows the following.

The "floating" wave P is available. It is excited from cycle to cycle at different times. This occurs very seldom. A study of the results of the automatic phase analysis shows the following: when the wave P is within the norm, then there is a PV2 deficit detectable – the deficit in the blood volume when blood enters the ventricles. The next cycle is a compensation one, and just in this cycle a shift of the wave P is detected. The Antecedent Anamnesis states that the patient suffers from a congenital heart defect – the defect of ventricular septum. It is evident from the ECG curve that there is a complete passivity of its contraction function. But it is impossible to detect, according to the ECG curve, the complete third degree block.

The weakness of function of the lungs is registered. This is confirmed by the modulation of the RHEO by breathing wave.

The therapy target could be normalization of the function of the lungs.

Example No. 15 from Practice

Antecedent Anamnesis
Patient (full name: XX15), female, born 1929. Since 2005 registered as tuberculous patient. Post-infarction cardioscle-rosis registered. Essential hypertension, degree II; pancreatic diabetes, type II. In childhood suffered from the left-side exudative pleurisy.

Diagnosis
Infiltrative pulmonary tuberculosis of upper lobe of the right lung, IIB, TMB–.

Clinical Examination
Complaints: dyspnea at physical activity, cough accompanied by sputum, weakness, hyperhidrosis.

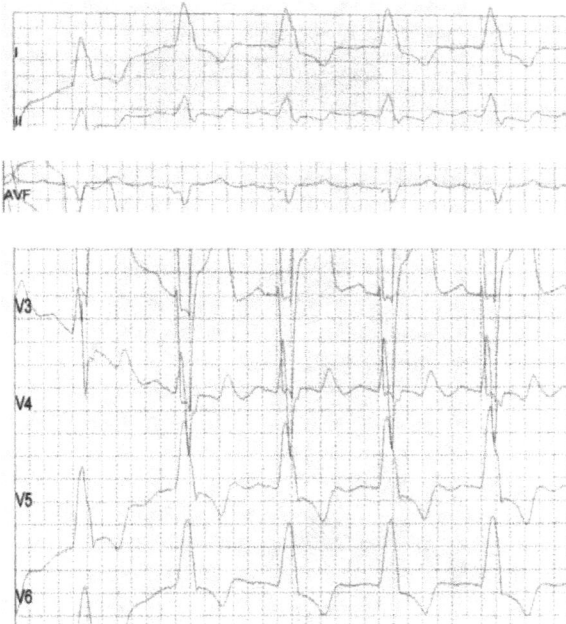

Basic data: weight – 60 kgs, stature – 162 cm; body mass index (BMI) – 22.86 kg/m2 (body mass norm).
Dynamic data: arterial pressure (AP) 140/90, 160/100.
Heart frequency: 68 beats per minute, breathing frequency: 19–21 breath movements per minute.
Auscultation: heart tones are damped down and rhythmic.
Breathing in lungs: rough, dry rales throughout the lungs. Abdomen: soft, no pains available. The liver borders are at the edge of the costal margin.

Tests

Laboratory Testing.
General blood test: leukocytes – $7.9*10*9$, erythrocytes – $4.52*10*12$, hemoglobin 100 g/l, erythrocyte sedimentation rate (ESR) – 3 mm/h, prothrombin – 105%, glucose – 5.22; bilirubin, total – 6.5, bilirubin conjugated – 0.0; AST – 0.35 μmole/l.h, ALT – 0.35 μmole/l.h, bacterial inoculation – no growth of tuberculosis mycobacteria. Urine analysis: protein traces, individual leukocytes. Instrumentation testing. ECG: P-0.08; PQ – 0.12; QRS – 0.16; QRST – 0.43; R–R – 0.66 Heart frequency – 91 beats per minute. Moderate sinus tachycardia. Scarry changes in the lower and anteroseptal areas against the background of complete left bundle-branch block. Apparent changes in myocardium.

Radiographic examination:

Right side, upper lobe: nidi and foci without clear contours in S1–S2 with destruction. Left side: lateral sinus is sealed. The roots are thickened. The heart is expanded to the left, the aorta is sclerous.

Date of entry – 20.07.2005
Date of birth – 28.04.1929
Full name – XX15 20-07-2005 11-44-02

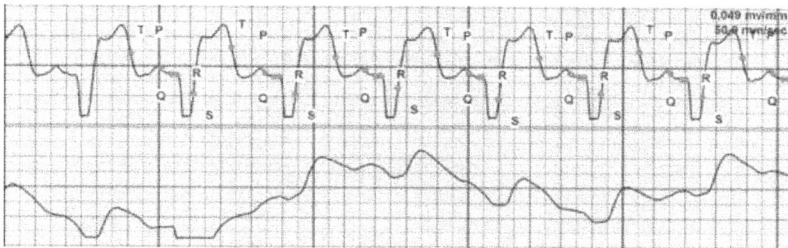

Diagnosis resulted from phase analysis.
SV (ml.l) – A dangerous increase in blood pressure in lesser circulation is available. MV (l) – A dangerous increase in blood pressure in greater circulation is available. The parameter PV5 should be thoroughly evaluated characterizing the blood volume pumped by the aorta as peristaltic pump. Pathological state.
PV1 (l) – An increased loading by volume of blood entering the left ventricle. Possible is an increase in blood pressure in lesser circulation. This measured parameter depends on the functional state of the lungs. The function of the lungs should be thoroughly studied with other instrumentation beyond the scope of supply of this analyzer. Pathological state. PV2 (l) – A dangerous increase in blood volume loading on the left atrium in atrial systole is available.
PV3 (l) – A dangerous increase in volume of blood ejected by the left ventricle in rapid ejection phase.
PV4 (l) – A dangerous increase in volume of blood ejected by the left ventricle in slow ejection phase.

PV5 (l) – A dangerous increase in volume of blood pumped by ascending aorta as peristaltic pump.

Doctor's diagnosis
The function of the aortic valve – the actual electric potential in the tension & opening phases of the valve shows a significant deviation from the normal values. There are several reasons responsible for this deviation. Every reason is based on certain changes in the actual heart anatomy in the following phases: the contraction of the ventricular septum, the contraction of the myocardium, the tension phase and the opening phase of the valve. The actual features in the tension phase are significant because the wave S is bifurcated, or more specifically, there are even three waves of "tinkling". This is caused by a pathology of myocardium (its apex). The RHEO shows that the pressure is increasing very slowly after the valve opening at the beginning of the phase, and a normal increase in pressure is started only in the middle of the phase. Then, the wave T undertakes its pumping function in proper manner. This provides the circulation of blood in blood vessels because the heart is practically not able to perform its functions (s. diagnosis resulted from the phase analysis).

Anatomic peculiarities of aortic valve in the tension phase indicated on the ECG curve. An increase in pressure in the aorta is started in the middle of the phase. There is a great amplitude of the SL wave half available.

Anatomic peculiarities of the aortic valve in the full opening phase indicated on the ECG curve. An increase in pressure is below the normal values.

Flabbiness of aorta – not available.

Condition of elasticity of aorta – norm.

The apex on the rheogram curve is parabolic and is positioned left referred to the wave point T on the ECG curve. The wave T is actually responsible for provision of blood flow in the blood vessel system.

Condition of contraction function of ventricular septum – considerable diminished. The R wave amplitude is at the level of the ECG isometric line.

Condition of contraction function of myocardium – the wave S is bifurcated. This is an indication of a significant pathology of myocardium. The duration of the QRS complex exceeds the normal values. This is responsible for abnormality in hemodynamics and for abnormal pathologic values (s. diagnosis resulted from the phase analysis).

Condition of venous blood flow – norm.

Condition of function of the lungs – norm.

Potential stroke problem – no sudden hemopulses are available. Potential stroke problem is not available.

ORTHOSTATIC TEST WAS NOT PERFORMED.

SV (ml) – Stroke volume

| SV(ml)=254.93 %=159.07 | 42.86 | | 98.40 |

MV (l) – Minute volume

| MV(l)=23.38 %=159.07 | 3.93 | | 9.02 |

PV1 (ml) – Volume of blood in phase of rapid and slow filling of left (aortic) ventricle of heart

| PV1(ml)=93.81 %=92.77 | 21.56 | | 48.66 |

PV2 (ml) – Volume of blood flowing into left (aortic) ventricle of heart during systole

| PV2(ml)=161.12 %=223.94 | 21.30 | | 48.74 |

PV3 (ml) – Volume of blood ejected by left (aortic) ventricle of heart during rapid ejection phase

| PV3(ml)=151.68 %=159.43 | 25.42 | | 58.46 |

PV4 (ml) – Volume of blood ejected by left (aortic) ventricle of heart during slow ejection phase

| PV4(ml)=103.25 %=158.54 | 17.44 | | 39.94 |

PV5 (ml) – Volume of blood transferred by ascending
aorta functioning as peristaltic pump

| PV5(ml)=23.72 %=107.69 | 6.89 | | 11.42 |

	QRS	RS	QT	PQ	TT	SV (ml)	MV (1)	PV1 (ml)	PV2 (ml)	PV3 (ml)	PV4 (ml)	PV5 (ml)
Min	0.080	0.035	0.324	0.069	0.654	42.9	3.9	21.6	21.3	25.4	17.4	6.9
Max	0.110	0.060	0.324	0.069	0.654	98.4	9.0	48.7	49.7	58.5	39.9	11.4
Average	0.137	0.106	0.391	0.065	0.654	254.9	23.4	93.8	161.1	151.7	103.3	23.7
0	0.139	0.105	0.387	0.067	0.653	250.6	23.0	90.6	160.0	149.1	101.5	23.0
1	0.139	0.105	0.400	0.055	0.674	256.8	22.9	118.3	138.5	152.8	104.0	24.3
2	0.143	0.109	0.396	0.067	0.657	268.0	24.5	92.3	175.7	159.5	108.5	24.3
3	0.135	0.105	0.396	0.059	0.661	256.8	23.3	107.8	149.0	152.8	104.0	24.3
4	0.139	0.109	0.387	0.055	0.648	265.7	24.6	114.3	151.4	158.1	107.6	23.8
5	0.135	0.105	0.392	0.072	0.665	254.8	23.0	96.1	158.7	151.6	103.2	23.9
6	0.135	0.101	0.392	0.067	0.657	239.5	21.9	86.6	153.0	142.5	97.0	23.0
7	0.135	0.101	0.396	0.067	0.669	241.4	21.6	94.7	146.7	143.6	97.8	23.4
8	0.135	0.101	0.392	0.067	0.653	239.5	22.0	82.5	157.1	142.5	97.0	23.0
9	0.139	0.105	0.392	0.072	0.661	252.7	22.9	89.3	163.4	150.4	102.3	23.5
10	0.143	0.114	0.392	0.051	0.653	281.1	25.8	128.1	153.0	167.3	113.8	24.6
11	0.131	0.101	0.387	0.072	0.644	239.5	22.3	73.3	166.2	142.5	97.0	23.0
12	0.135	0.105	0.392	0.072	0.644	254.8	23.7	72.6	182.2	151.6	103.2	23.9
13	0.139	0.109	0.379	0.072	0.627	261.1	25.0	68.4	192.7	155.4	105.7	22.9
14	0.131	0.101	0.383	0.063	0.644	237.6	22.1	89.9	147.7	141.4	96.3	22.6
15	0.135	0.105	0.392	0.059	0.648	254.8	23.6	98.9	155.9	151.6	103.2	23.9
16	0.131	0.101	0.396	0.072	0.653	243.3	22.4	74.5	168.8	144.7	98.6	23.8
17	0.131	0.097	0.396	0.067	0.648	228.0	21.1	69.9	158.2	135.6	92.4	22.9
18	0.139	0.109	0.396	0.072	0.653	270.2	24.8	82.7	187.5	160.8	109.4	24.7
19	0.131	0.105	0.383	0.063	0.644	252.7	23.5	98.0	154.7	150.4	102.3	23.5
20	0.143	0.109	0.387	0.055	0.653	263.4	24.2	114.5	148.9	156.8	106.7	23.4
21	0.143	0.109	0.400	0.067	0.674	270.2	24.1	106.0	164.2	160.8	109.4	24.7
22	0.139	0.105	0.392	0.067	0.653	252.7	23.2	87.0	165.7	150.4	102.3	23.5
23	0.143	0.114	0.392	0.072	0.665	281.1	25.4	106.0	175.1	167.3	113.8	24.6

Comparative analysis of example No. 15 from practice

Phase duration.

The QRS complexes are equal in their duration.

A difference in the phase duration between the PQ phases is 0.08s.

A difference in the phase duration between the QT phases is 0.077s.

Diagnoses.

Classical diagnosis option: Scarry changes in the lower and antero-septal area against the background of complete left bundle-branch block. Apparent changes in myocardium.

The phase analysis shows the following.

The values obtained from the automatic phase analysis show that this patient must be carefully monitored in a hospital. This is required due to the fact that there are significant changes in his heart anatomy. This is also noted in part 1 of the Doctor's Diagnosis above.

Seven months later the patient died. The declared cause of death is stroke. But at the moment of the testing & examination described above no sudden AP pulses were detected. It should be mentioned that an orthostatic test was not performed, too.

Example No. 16 from Practice

Antecedent Anamnesis
Patient (full name: XX16), male, born 1950. Since 2005 registered as tuberculous patient.

Diagnosis
Disseminated pulmonary tuberculosis, infiltration & destruction phase IA, TMB +.

Clinical Examination
Complaints: weakness, cough accompanied by sputum hardly discharged; dyspnea at minor physical activity, low-temperature fever. Basic data: weight – 60 kgs, stature – 175 cm; body mass index (BMI) – 19.39 kg/m2 (body mass deficit).

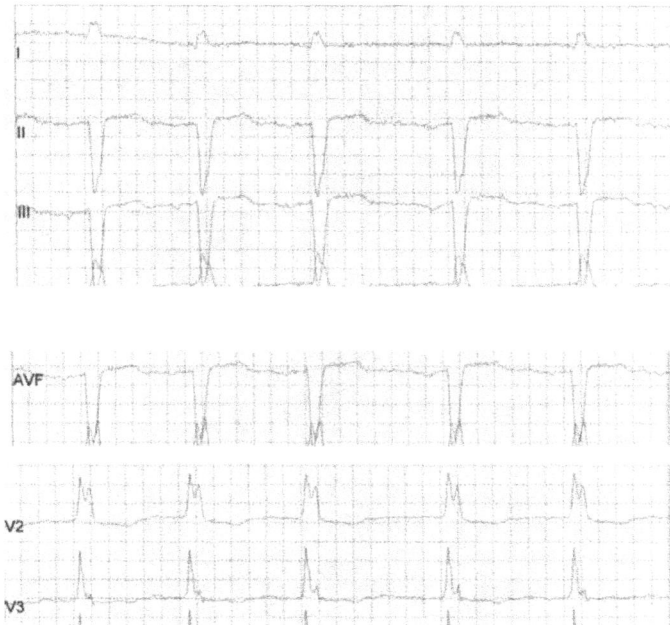

Dynamic data: arterial pressure (AP) 120/70, 140/80.
Heart frequency: 97 beats per minute, breathing frequency: 19–20 breath movements per minute.
Auscultation: heart tones are clear and rhythmic.
Breathing in lungs: vesicular. Abdomen: soft, no pains available. The liver borders are at the edge of the costal margin.

Tests
Laboratory Testing.
General blood test: leukocytes – 5.4*10*9, erythrocytes – 4.72*10*12, hemoglobin 124 g/l, erythrocyte sedimentation rate (ESR) – 40 mm/h, prothrombin – 114%, glucose – 5.0; bilirubin, total – 5.1, bilirubin conjugated – 0.0; AST – 0.17 μmole/l.h, ALT – 0.17 μmole/l.h, bacterial inoculation – rich growth of tuberculosis mycobacteria. Urine analysis: no pathology.

Instrumentation testing. ECG: PQ – ff; QRS – 0.13; QRST – 0.41; R–R – 0.60–0.80.
Heart frequency – 100–75 beats per minute. Atrial fibrillation accompanied by ventricular arrhythmia; ventricle contraction frequency is 100-75 beats per minute. Complete right bundle-branch block; left anterior fascicular block. Scarry changes in the lower wall of the left ventricle, coronary insufficiency within the area of scarry changes. Apparent changes in myocardium.

Radiographic examination:

Right side, throughout the length: nidi without clear contours available against the background of ring fibrosis; in the middle portion – fused. Left side, below rib II: nidi without clear contours, in the middle portion – inhomogeneous infiltrate with a destruction cavity of 2.0 cm. The roots are expanded. The heart is expanded crosswise.

Date of entry – 20.07.2005
Date of birth – 25.04.1950
Full name – XX16 20-07-2005 9-56-46

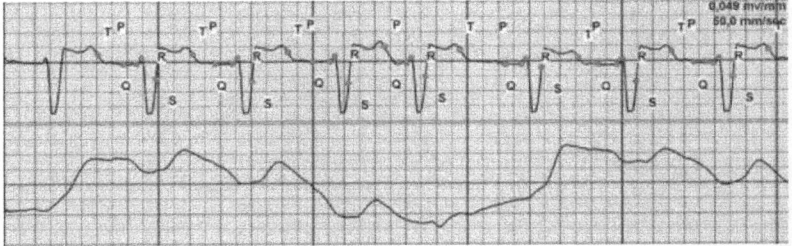

Diagnosis resulted from phase analysis.
SV (ml.l) – A dangerous increase in blood pressure in lesser circulation is available. MV (l) – A dangerous increase in blood pressure in lesser circulation is available. The parameter PV5 should be thoroughly evaluated characterizing the blood volume pumped by the aorta as peristaltic pump. Pathological state.
PV1 (l) – A dangerous increase in loading by volume of blood entering the left ventricle. PV2 (l) – A dangerous increase in blood volume loading on the left atrium in atrial systole. PV3 (l) – A dangerous increase in volume of blood ejected by the left ventricle in rapid ejection phase.
PV4 (l) – A dangerous increase in volume of blood ejected by the left ventricle in slow ejection phase.
PV5 (l) – A dangerous increase in volume of blood pumped by ascending aorta as peristaltic pump.

Doctor's diagnosis.
The function of the aortic valve – despite the fact that there are great deviations from the normal hemodynamic parameters, the function of the aortic valve is normal (s. diagnostics based on the phase analysis).

Anatomic peculiarities of aortic valve in the tension phase indicated on the ECG curve – norm.

Anatomic peculiarities of the aortic valve in the full opening phase indicated on the ECG curve. A changed condition as an increase in the Lj wave amplitude is available.

Flabbiness of aorta – in cycle 4 and cycle 5 available is a sharp decrease in the maximum pressure in slow ejection phase. This is determined by breathing process which modulates the AP in the aorta. This is a pathology. Other cycles do not show such deviations.

Condition of elasticity of aorta – norm.

The apex on the rheogram curve is parabolic and is positioned left referred to the wave

point T on the ECG curve.

Condition of contraction function of ventricular septum – considerable diminished. The R wave amplitude is less than that of the T wave.

Condition of contraction function of myocardium – there is a pathology as a bifurcation of the wave S available. It is responsible for significant deviations in hemodynamics from the relevant norms (s. diagnosis resulted from the phase analysis).

Condition of venous blood flow – norm.

Condition of function of the lungs – the function of the lungs is diminished. A minor modulation of the rheogram curve by breathing is available.

Potential stroke problem – no sudden hemopulses are available. Potential stroke problem is not available.

ORTHOSTATIC TEST WAS NOT PERFORMED.

SV (ml) – Stroke volume

| SV(ml)=203.73 %=123.92 | 40.74 | 90.98 |

MV (l) – Minute volume

| MV(l)=20.08 %=123.92 | 4.02 | 8.97 |

PV1 (ml) – Volume of blood in phase of rapid and slow filling of left (aortic) ventricle of heart

| PV1(ml)=-2.94 %=114.00 | 20.27 | 44.47 |

PV2 (ml) – Volume of blood flowing into left (aortic) ventricle of heart during systole

| PV2(ml)=206.57 %=344.13 | 20.47 | 46.51 |

PV3 (ml) – Volume of blood ejected by left (aortic) ventricle of heart during rapid ejection phase

| PV3(ml)=121.20 %=124.07 | 24.17 | 54.09 |

PV4 (ml) – Volume of blood ejected by left (aortic) ventricle of heart during slow ejection phase

| PV4(ml)=82.52 %=123.89 | 16.57 | 36.89 |

PV5 (ml) – Volume of blood transferred by ascending
aorta functioning as peristaltic pump

| PV5(ml)=19.24 %=100.95 | 6.11 | 9.57 |

	QRS	RS	QT	PQ	TT	SV (ml)	MV (1)	PV1 (ml)	PV2 (ml)	PV3 (ml)	PV4 (ml)	PV5 (ml)
Min	0.080	0.035	0.289	0.065	0.609	40.7	4.0	20.3	20.5	24.2	16.6	6.1
Max	0.110	0.060	0.289	0.065	0.609	91.0	9.0	44.5	46.5	54.1	36.9	9.6
Average	0.126	0.094	0.358	0.107	0.609	203.7	20.1	-2.8	206.6	121.2	82.5	19.2
0	0.126	0.093	0.362	0.084	0.614	201.1	19.6	37.9	163.2	119.6	81.5	19.3
1	0.126	0.093	0.362	0.109	0.618	201.1	19.5	3.9	197.1	119.6	81.5	19.3
2	0.122	0.088	0.353	0.109	0.623	185.7	17.9	21.2	164.5	110.5	75.3	18.2
3	0.122	0.093	0.345	0.050	0.492	195.7	23.9	-837.1	1032.8	116.4	79.3	18.2
4	0.130	0.101	0.370	0.105	0.766	231.2	18.1	118.5	112.8	137.6	93.7	21.3
5	0.126	0.093	0.353	0.118	0.602	197.5	19.7	-29.5	227.0	117.5	80.0	18.6
6	0.126	0.093	0.353	0.126	0.627	197.5	18.9	3.5	194.0	117.5	80.0	18.6
7	0.130	0.101	0.366	0.105	0.635	229.2	21.7	33.6	195.6	136.4	92.8	20.9
8	0.130	0.097	0.366	0.122	0.618	215.0	20.9	-31.3	246.3	127.9	87.1	20.1
9	0.126	0.097	0.366	0.122	0.618	216.9	21.0	-32.3	249.2	129.0	87.9	20.5
10	0.118	0.084	0.362	0.118	0.610	176.7	17.4	-26.4	203.1	105.1	71.7	18.3
11	0.126	0.093	0.358	0.139	0.602	199.3	19.9	-104.2	303.5	118.6	80.7	19.0
12	0.130	0.101	0.358	0.105	0.610	225.0	22.1	4.6	220.4	133.9	91.1	20.1
13	0.126	0.093	0.366	0.122	0.606	202.8	20.1	-65.8	268.6	120.6	82.2	19.7
14	0.130	0.097	0.353	0.118	0.585	209.2	21.5	-86.6	295.8	124.5	84.7	18.9
15	0.122	0.093	0.353	0.076	0.606	199.3	19.7	50.7	148.6	118.6	80.7	19.0
16	0.126	0.093	0.353	0.105	0.597	197.5	19.8	-12.1	209.6	117.5	80.0	18.6
17	0.126	0.097	0.353	0.101	0.597	211.2	21.2	-4.1	215.3	125.6	85.5	19.3
18	0.130	0.097	0.366	0.122	0.610	215.0	21.2	-55.4	270.4	127.9	87.1	20.1
19	0.126	0.093	0.358	0.114	0.593	199.3	20.2	-54.3	253.6	118.6	80.7	19.0
20	0.135	0.101	0.349	0.067	0.580	218.5	22.6	39.2	179.3	130.1	88.4	18.8
21	0.122	0.093	0.353	0.118	0.610	199.3	19.6	-11.6	210.9	118.6	80.7	19.0
22	0.126	0.093	0.353	0.122	0.597	197.5	19.8	-50.9	248.4	117.5	80.0	18.6
23	0.122	0.088	0.362	0.072	0.610	188.9	18.6	48.1	140.8	112.4	76.6	18.9
24	0.118	0.088	0.349	0.122	0.593	185.7	18.8	-49.2	234.9	110.5	75.3	18.2
25	0.126	0.093	0.362	0.118	0.606	201.1	19.9	-40.9	242.0	119.6	81.5	19.3

Comparative analysis of example No. 16 from practice

Phase duration.

The QRS complexes are equal in their duration.

A difference in the phase duration between the QT phases is 0.052s.

Diagnoses.

Classical diagnosis option: complete right bundle-branch block; left anterior fascicular block. Scarry changes in the lower wall of the left ventricle.

The phase analysis shows the following. The wave P is not available. The amplitude of the wave R is a minimum. The wave S is extended. This complete set of diagnostics data is an indication of total functional abnormality of all heart segments. In this case, only the aorta functions supporting the structure of blood circulation as a whole. But the hemodynamic parameters are extremely critical (s. deviations from the hemodynamic values).

In this case, therapy in a hospital should be recommended under special monitoring. Further patient data are not available.

Example No. 17 from Practice

Antecedent Anamnesis
Patient (full name: XX17), male, born 1932. Since 2005 registered as tuberculous patient.

Diagnosis
Nidal tuberculosis S1-S2 of both lungs, infiltration phase, IA, TMB –. Coronary heart disease, postinfarction cardiosclerosis with rhythm abnormality (ciliary arrhythmia), N2.

Clinical Examination
Complaints: weakness and dyspnea at physical activity.
Basic data: weight – 66 kgs, stature – 169 cm; body mass index (BMI) – 22.9 kg/m2 (body mass norm).
Dynamic data: arterial pressure (AP) 180/100, 140/80.
Heart frequency: 97–108 beats per minute, breathing frequency: 19–20 breath movements per minute.
Auscultation: muffled heart tones; ciliary arrhythmia.

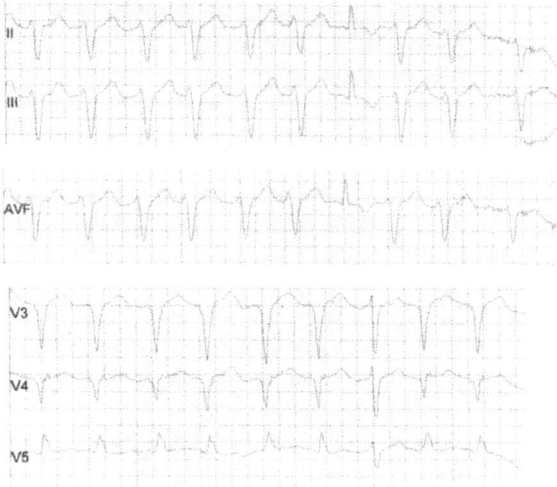

Breathing in lungs: rough; dry sibilant rales.
Abdomen: soft, no pains available. The liver borders extend 2.5 cm beyond the costal margin.

Tests
Laboratory Testing.
General blood test: leukocytes – 5.4*10*9, erythrocytes – 4.72*10*12, hemoglobin 124 g/l, erythrocyte sedimentation rate (ESR) – 15 mm/h, prothrombin – 114%, glucose – 4.3, bilirubin, total – 5.8, bilirubin conjugated – 0.0; AST – 0.33 μmole/l.h, ALT – 0.17 μmole/l.h. Urine analysis: no pathology.

Instrumentation testing.
ECG: PQ – ff; QRS – 0.11; QRST – 0.28; R–R – 0.32–0.40.
Heart frequency – 187–150 beats per minute.

Atrial fibrillation with ventricular arrhythmia; contraction frequency for ventricles – 187–150 beats per minute; single ventricular extrasystoles. Left anterior fascicular block. Scarry changes in anteroseptal area. Myocardial hyper-throphy of left ventricle. Repolarization disturbances in the anteroseptal area of coronary insufficiency type, moderate repolarization disturbances in the lateral wall of the left ventricle.

Radiographic examination:

On both sides, throughout the length: apparent diffuse pulmonary fibrosis in S1–S2 on both sides: nidi and foci up to 1.5 cm without clear contours. The roots are fibrous. The heart is expanded crosswise towards the left.

Date of entry – 22.06.2005
Date of birth – 08.04.1932
Full name – XX17 22-06-2005 10-27-33

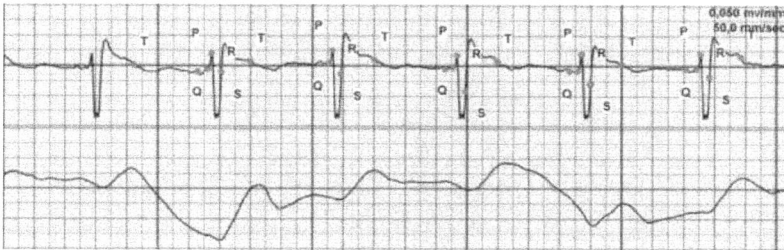

Diagnosis resulted from phase analysis.
PV1 (l) – An increased loading by volume of blood entering the left ventricle is available. Possible is an increase in blood pressure in lesser circulation. This measured parameter depends on the functional state of the lungs. Therefore, the function of the lungs should

be thoroughly studied with other instrumentation beyond the scope of supply of this analyzer.
The volume-related parameters of function of the heart and the associated blood vessels are within normal values.

Doctor's diagnosis.
The function of the aortic valve – norm.

Anatomic peculiarities of aortic valve in the tension phase indicated on the ECG curve. A changed condition as an increase in the SL wave amplitude is available.

Anatomic peculiarities of the aortic valve in the full opening phase indicated on the ECG curve. A changed condition as the smoothed wave Lj is available.

Flabbiness of aorta – not available.
Condition of elasticity of aorta – norm.
The rheogram apex is parabolic and is positioned left of the point T on the ECG curve.

Condition of contraction function of ventricular septum – considerably diminished. The amplitude of the apex R is lower than the amplitude of the wave T.

Condition of contraction function of myocardium - the function is considerably diminished. The amplitude of the lower portion of the QRS complex is bifurcated, and that is an indication of a pathology of myocardium, precisely, its apex.

Condition of venous blood flow. - Hindered.

Condition of function of the lungs - the function of the lungs is diminished. There is a minor modulation of the rheogram curve by breathing available.

Potential stroke problem - no sudden hemopulses are available. Potential stroke problem is not available.

ORTHOSTATIC TEST WAS NOT PERFORMED.

Other features. Due to a weak potential of the wave P on the ECG curve, the parameter PV1 is slightly greater, but it is at the top limit of the norm.

SV (ml) – Stroke volume
SV(ml)=92.07 %=0.00   43.21   99.62

MV (l) – Minute volume
MV(l)=6.95 %=0.00   3.28   7.52

PV1 (ml) – Volume of blood in phase of rapid and slow filling of left (aortic) ventricle of heart
PV1(ml)=68.26 %=10.94   27.00   61.53

PV2 (ml) – Volume of blood flowing into left (aortic) ventricle of heart during systole
PV2(ml)=23.81 %=0.00   16.20   38.09

PV3 (ml) – Volume of blood ejected by left (aortic) ventricle of heart during rapid ejection phase
PV3(ml)=54.71 %=0.00   25.62   59.18

PV4 (ml) – Volume of blood ejected by left (aortic) ventricle of heart during slow ejection phase
PV4(ml)=37.37 %=0.00   17.58   40.44

PV5 (ml) – Volume of blood transferred by ascending
aorta functioning as peristaltic pump
PV5(ml)=10.70 %=0.00   7.02   11.75

	QRS	RS	QT	PQ	TT	SV (ml)	MV (1)	PVl (ml)	PV2 (ml)	PV3 (ml)	PV4 (ml)	PV5 (ml)
Min	0.080	0.035	0.330	0.079	0.795	43.2	3.3	27.0	16.2	25.6	17.6	7.0
Max	0.110	0.060	0.330	0.079	0.795	99.6	7.5	61.5	38.1	59.2	40.4	11.7
Average	0.083	0.058	0.290	0.054	0.795	92.1	7.0	68.3	23.8	54.7	37.4	10.7
0	0.088	0.063	0.278	0.063	0.753	101.0	8.0	71.0	29.9	60.0	40.9	10.7
1	0.084	0.059	0.299	0.050	0.803	95.7	7.1	71.9	23.9	56.9	38.9	11.3
2	0.080	0.055	0.290	0.050	0.795	84.4	6.4	63.4	21.1	50.1	34.3	10.2
3	0.084	0.059	0.286	0.050	0.803	93.3	7.0	70.5	22.8	55.4	37.9	10.6
4	0.084	0.059	0.286	0.067	0.770	93.3	7.3	65.1	28.2	55.4	37.9	10.6
5	0.080	0.055	0.303	0.050	0.812	86.4	6.4	65.0	21.4	51.3	35.1	10.8
6	0.088	0.063	0.290	0.050	0.791	103.9	7.9	77.8	26.1	61.8	42.2	11.4
7	0.080	0.055	0.299	0.067	0.791	85.8	6.5	60.2	25.6	50.9	34.8	10.6
8	0.084	0.063	0.299	0.050	0.774	106.7	8.3	79.7	27.0	63.4	43.3	12.1
9	0.084	0.059	0.303	0.046	0.808	96.5	7.2	73.5	23.0	57.3	39.2	11.5
10	0.084	0.059	0.282	0.050	0.799	92.5	6.9	69.9	22.6	54.9	37.5	10.4

Comparative analysis of example No. 17 from practice

Phase duration.

A difference in the duration of the QRS complexes is 0.027 s.

The QT phases are equal in their duration.

Diagnoses.

Classical diagnosis option: Left anterior fascicular block. Scarry changes in anter-oseptal area.

The phase analysis shows the following.

The wave P is not available. The wave S is extended. The wave T manifests itself as a weak wave, but the hemodynamic parameters are within the norm.

The patient's arterial pressure is very high. Taking into account the patient's age and his disease, the therapy target should be recommended as follows: improvement in lung function because hemodynamic parameters are already within the norm.

Example No. 18 from Practice

Antecedent Anamnesis
Patient (full name: XX18), male, born 1948. Since 2005 registered as tuberculous patient, pathological changes in lungs detected during preventive examination.

Diagnosis
Infiltrative tuberculosis S1-S2 of both lungs, IA, TMB –.
Coronary heart disease, essential hypertension.

Clinical Examination
Complaints: weakness, dry cough every morning.
Basic data: weight – 66 kgs, stature – 175 cm;
body mass index (BMI) – 21.35 kg/m2 (body mass norm).
Dynamic data: arterial pressure (AP) 16 0/110 , 14 0/9 0.
Heart frequency: 75 beats per minute, breathing frequency: 19–20 breath movements per minute.
Auscultation: heart tones are clear and rhythmic.
Breathing in lungs: vesicular. Abdomen: soft, no pains available.
The liver borders: the liver is not increased.

Tests
Laboratory Testing.
General blood test: leukocytes – 6.1*10*9, erythrocytes – 4.72*10*12, hemoglobin 123 g/l, erythrocyte sedimentation rate (ESR) – 6 mm/h, prothrombin – 114%, glucose – 3.9; bilirubin, total – 11.7, bilirubin conjugated – 0.0; AST – 0.17μmole/l.h, ALT – 0.42 μmole/l.h. Urine analysis: no pathology.

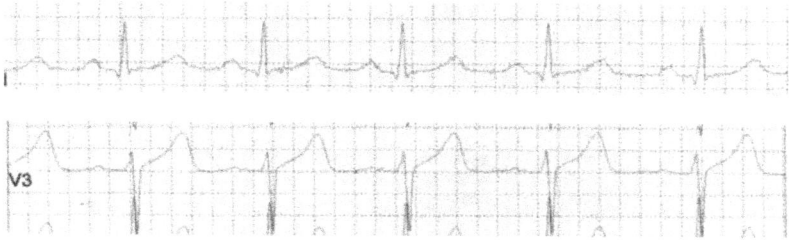

Instrumentation testing.
ECG: P – 0.08; PQ – 0.16; QRS – 0.08;
QRST – 0.38; R–R – 0.69
Heart frequency – 87 beats per minute.
Moderate sinus tachycardia. Myocardial hypertrophy of the right atrium. Diffuse changes in myocardium.

Radiographic examination:

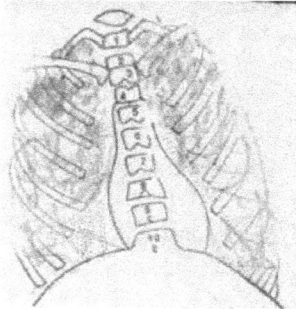

On both sides in S1–S2: multiple nidi without clear contours which are clustered in S2 as foci. The roots are fibrous. The heart is extended transversely to the left.

Date of entry – 04.07.2005
Date of birth – 31.03.1948
Full name – XX18 04-07-2005 9-43-36

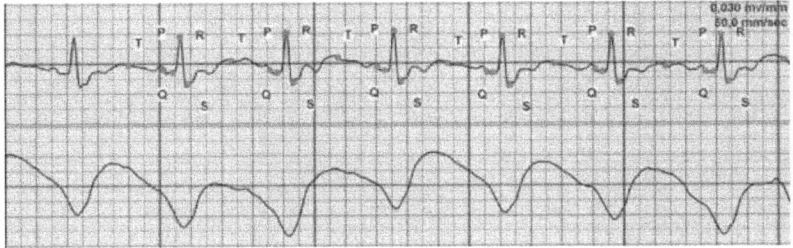

Diagnosis resulted from phase analysis.
PV5 (l) – Within the norm.
The volume-related parameters of function of the heart and the associated blood vessels
are within normal values.

Doctor's diagnosis.
The function of the aortic valve – considerably diminished. There is a contraction of the ventricular septum even in the phase QR available, and next there is also a contraction of myocardium in the phase RS available, the valve is leaking blood into the aorta, and that is the cause of an increase in a minimum value of the AP.

Anatomic peculiarities of aortic valve in the tension phase indicated on the ECG curve. A changed condition as the smoothed SL wave is available.

Anatomic peculiarities of the aortic valve in the full opening phase indicated on the ECG curve. A changed condition as an increase in the wave Lj amplitude and its considerable extension is available.

Flabbiness of aorta – not available.

Condition of elasticity of aorta – slightly diminished. The rheogram apex is flat.

Condition of contraction function of ventricular septum – norm.

Condition of contraction function of myocardium – the function is considerably diminished. The amplitude of the lower portion of the QRS complex is a minimum, and that is an indication of the maximum passivity of the myocardium contraction.

Condition of venous blood flow – norm.

Condition of function of the lungs –norm.

Potential stroke problem – no hemopulses are available.

ORTHOSTATIC TEST WAS NOT PERFORMED.

SV (ml) – Stroke volume
SV(ml)=80.46 %=0.00   42.26   96.32

MV (l) – Minute volume
MV(l)=6.74 %=0.00   3.54   8.07

PV1 (ml) – Volume of blood in phase of rapid and slow filling of left (aortic) ventricle of heart
PV1(ml)=41.51 %=0.00   24.72   55.58

PV2 (ml) – Volume of blood flowing into left (aortic) ventricle of heart during systole
PV2(ml)=38.95 %=0.00   17.54   40.74

PV3 (ml) – Volume of blood ejected by left (aortic) ventricle of heart during rapid ejection phase
PV3(ml)=47.73 %=0.00   25.07   57.24

PV4 (ml) – Volume of blood ejected by left (aortic) ventricle of heart during slow ejection phase
PV4(ml)=32.73 %=0.00   17.20   39.08

PV5 (ml) – Volume of blood transferred by ascending
aorta functioning as peristaltic pump
PV5(ml)=12.26 %=12.61   6.66   10.88

	QRS	RS	QT	PQ	TT	SV (ml)	MV (1)	PV1 (ml)	PV2 (ml)	PV3 (ml)	PV4 (ml)	PV5 (ml)
Min	0.080	0.035	0.313	0.074	0.716	42.3	3.5	24.7	17.5	25.1	17.2	6.7
Max	0.110	0.060	0.313	0.074	0.716	96.3	8.1	55.6	40.7	57.2	39.1	10.9
Average	0.104	0.049	0.406	0.051	0.716	80.5	6.7	41.5	38.9	47.7	32.7	12.3
0	0.097	0.050	0.374	0.059	0.660	82.3	7.5	36.8	45.5	48.8	33.5	11.9
1	0.109	0.046	0.395	0.046	0.693	72.0	6.2	35.9	36.1	42.7	29.3	11.0
2	0.105	0.046	0.408	0.050	0.710	73.3	6.2	36.5	36.8	43.5	29.8	11.5
3	0.109	0.050	0.395	0.042	0.685	83.1	7.3	42.2	41.0	49.3	33.8	12.2
4	0.109	0.050	0.408	0.046	0.719	84.4	7.0	44.7	39.7	50.1	34.3	12.6
5	0.097	0.050	0.454	0.055	0.769	89.5	7.0	47.3	42.2	53.1	36.4	14.5
6	0.109	0.046	0.395	0.042	0.630	72.0	6.8	24.5	47.5	42.7	29.3	11.0
7	0.105	0.046	0.408	0.046	0.723	73.3	6.1	39.3	34.0	43.5	29.8	11.5
8	0.105	0.050	0.395	0.050	0.710	83.6	7.1	44.1	39.4	49.6	34.0	12.3
9	0.097	0.046	0.391	0.059	0.719	72.6	6.1	38.2	34.4	43.1	29.6	11.2
10	0.097	0.050	0.357	0.059	0.681	80.5	7.1	42.4	38.1	47.7	32.7	11.3
11	0.101	0.046	0.412	0.038	0.765	74.0	5.8	46.2	27.8	43.9	30.1	11.7
12	0.097	0.050	0.387	0.059	0.693	83.6	7.2	41.4	42.1	49.6	34.0	12.3
13	0.113	0.050	0.412	0.046	0.740	84.4	6.8	46.2	38.1	50.1	34.3	12.6
14	0.109	0.046	0.408	0.046	0.727	73.0	6.0	39.1	33.9	43.3	29.7	11.3
15	0.105	0.046	0.471	0.050	0.799	77.8	5.8	42.1	35.7	46.1	31.7	13.2
16	0.097	0.050	0.395	0.067	0.672	84.4	7.5	32.3	52.1	50.1	34.3	12.6
17	0.105	0.050	0.412	0.050	0.740	85.2	6.9	46.7	38.5	50.5	34.6	12.9
18	0.109	0.050	0.458	0.055	0.786	88.8	6.8	46.7	42.1	52.7	36.1	14.2
19	0.105	0.050	0.391	0.050	0.664	83.1	7.5	36.4	46.7	49.3	33.8	12.2
20	0.109	0.050	0.378	0.050	0.706	81.4	6.9	44.0	37.4	48.3	33.1	11.6
21	0.097	0.050	0.420	0.063	0.765	86.7	6.8	47.2	39.5	51.4	35.3	13.4

Comparative analysis of example No. 18 from practice

Phase duration.

A difference in the phase duration between the QRS complexes is 0.024s.

A difference in the phase duration between the PQ phases is 0.109s. A difference in the phase duration between the QT phases is 0.026s. Diagnoses.

Classical diagnosis option: Myocardial hypertrophy of the right atrium. Diffuse changes in myocardium.

The phase analysis shows the following.

The initial step of the visual examination of the relevant curves is to identify three factors of great importance. The first factor is the bell-type RHEO curve. This is indicative of a pathology of the aortic valve leaking blood in the phase of the ventricular contraction. This is the cause of an increase in the arterial pressure. The second factor is an inversion of the wave T. The actual shape of the wave T is resulted from the loading on the aorta, and that is registered in the diagnosis produced automatically. A deviation from the norm is in this case 12%. The third factor is the small wave S. Actually, there is no contraction of the ventricular walls available.

The lungs function is within the norm.

The therapy target should be an improvement in elasticity of the aorta and blood vessels.

Example No. 19 from Practice

Antecedent Anamnesis
Patient (full name: XX19), female, born 1965. Since 1997 registered as tuberculous patient, pulmonary tuberculosis.

Diagnosis
Pulmonary fibrous cavernous tuberculosis, infiltration & seeding phase. IIB, Tuberculosis Mycobacteria (TMB) +. Consequences of craniocerebral injury.

Clinical Examination
Complaints: dyspnea, cough with white sputum discharged, hyper-hidrosis, pains in right hypochondrium. Basic data: weight – 57 kgs, stature – 165 cm; body mass index (BMI) – 20.93 kg/m2 (body mass norm).

Dynamic data: arterial pressure (AP) 100/60, 110/70, 90/60. Heart frequency: 89 beats per minute, breathing frequency: 21–23 breath movements per minute.

Auscultation: heart tones are damped down and rhythmic.

Breathing: rough on both sides, dry rales on both sides in lungs.

Abdomen: pains in the right hypochon-drium.

Liver borders: the liver extends 2 cm beyond the costal margin.

Tests
Laboratory Testing.

General blood test: leukocytes – 8.1*10*9, erythrocytes – 4.32*10*12, hemoglobin 108 g/l, erythrocyte sedimentation rate (ESR) – 52 mm/h, prothrombin – 106%, glucose – 3.9 μmole/l; bilirubin, total – 10.9, bilirubin conjugated – 0.0; AST – 0.5 μmole/l.h, ALT – 2.7 μmole/l.h, bacterial inoculation of sputum – rich growth of tuberculosis mycobacteria. Urine analysis: leukocytes 12–15 in visible field, erythrocytes – 4–7 in visible field, bacteria – a considerable amount.

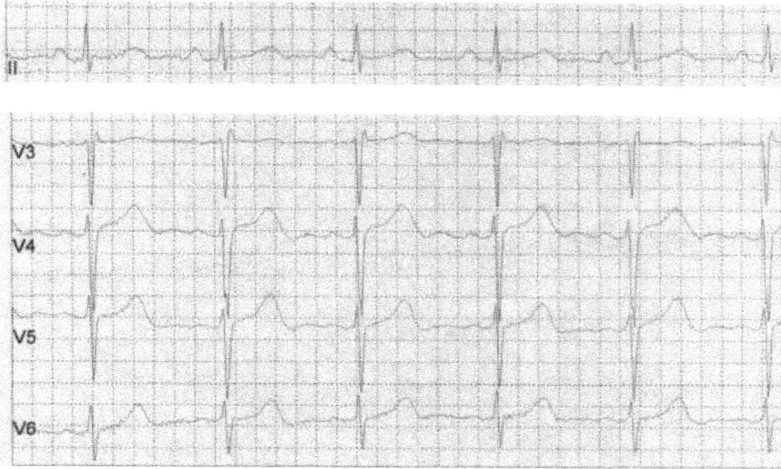

Instrumentation testing.
ECG: P – 0.10; PQ – 0.12; QRS – 0.08; QRST – 0.36; R–R – 0.67
Heart frequency – 89 beats per minute. Moderate sinus tachycardia. Myocardial hypertrophy of the right atrium. Incomplete right bundle-branch block. Myocardial hypertrophy of the right ventricle (S-type). Diffuse changes in myocardium.

Radiographic examination:

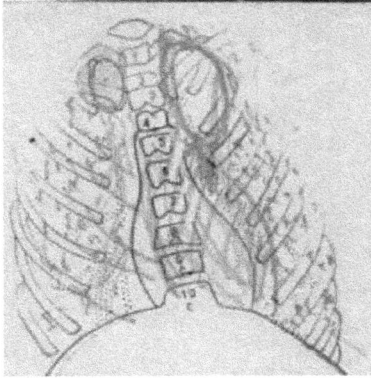

Right side, throughout the lung: against the background of emphysema, small-sized nidi without clear contours, apically a fibrous cavity with a diameter of 5 cm. Left side: in the upper lobe a cavity of 10*8 cm with fibrous walls, below the cavity – nidi without clear contours. Mediastinum organs are shifted to the left.

Date of entry – 03.06.2005
Date of birth – 27.09.1965
Full name – XX19 03-06-05 09-38-03

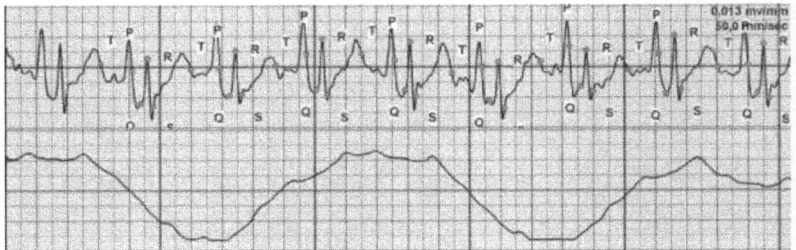

Diagnosis resulted from phase analysis.
Automatic diagnosis.
The volume-related parameters of function of the heart and the associated blood vessels are within normal values.

Doctor's diagnosis.
The function of the aortic valve – norm.

Anatomic peculiarities of aortic valve in the tension phase indicated on the ECG curve – norm.

Anatomic peculiarities of the aortic valve in the full opening phase indicated on the ECG curve – norm.

Flabbiness of aorta – not available.

Condition of elasticity of aorta – norm.

The rheogram apex is parabolic and is positioned left to the point T on the ECG curve.

Condition of contraction function of ventricular septum – norm.
Condition of contraction function of myocardium – norm.

The amplitude of the lower portion of the QRS complex is a minimum, and that is an indication of the maximum passivity of the myocardium contraction.

Condition of venous blood flow – norm.

Condition of function of the lungs – the function of the lungs is considerably diminished.

Potential stroke problem – no sudden hemopulses are available. No potential stroke problem is available.

ORTHOSTATIC TEST WAS NOT PERFORMED.

Additional information. Irrespective the fact that there is an increased amplitude of the wave T detected, the pumping function of the aorta is actually not available, and there is no increase in the pressure in this phase.

Of special note is the wave P. Its amplitude is greater than the QRS complex. The trailing edge of the pulse shows a significant dip, and this corresponds to a heavy pathology. The waves on the RHEO have a small amplitude. Ignoring all these facts, it should be stated that the hemodynamic parameters are within the norm (s. phase analysis).

SV (ml) – Stroke volume
SV(ml)=52.07 %=0.00   41.58   33.91

MV (l) – Minute volume
MV(l)=5.49 %=0.00   4.39   9.30

PV1 (ml) – Volume of blood in phase of rapid and slow filling of left (aortic) ventricle of heart
PV1(ml)=18.82 %=0.00   15.77   34.81

PV2 (ml) – Volume of blood flowing into left (aortic) ventricle of heart during systole
PV2(ml)=33.25 %=0.00   25.81   59.10

PV3 (ml) – Volume of blood ejected by left (aortic) ventricle of heart during rapid ejection phase
PV3(ml)=30.89 %=0.00   24.66   55.82

PV4 (ml) – Volume of blood ejected by left (aortic) ventricle of heart during slow ejection phase
PV4(ml)=21.18 %=0.00   16.91   38.09

PV5 (ml) – Volume of blood transferred by ascending
aorta functioning as peristaltic pump
PV5(ml)=7.66 %=0.00   6.41   10.28

	QRS	RS	QT	PQ	TT	SV (ml)	MV (1)	PV1 (ml)	PV2 (ml)	PV3 (ml)	PV4 (ml)	PV5 (ml)	HR
Min	0.080	0.035	0.302	0.062	0.569	41.6	4.4	15.8	25.8	24.7	16.9	6.4	105.5
Max	0.110	0.060	0.302	0.062	0.569	93.9	9.9	34.8	59.1	55.8	38.1	10.3	105.5
Average	0.087	0.040	0.316	0.055	0.569	52.1	5.5	18.8	33.3	30.9	21.2	7.7	105.5
0	0.084	0.038	0.324	0.055	0.585	48.8	5.0	19.2	29.5	28.9	19.8	7.5	102.5
1	0.080	0.042	0.303	0.063	0.552	56.6	6.2	17.4	39.2	33.6	23.0	8.0	108.8
2	0.076	0.038	0.295	0.063	0.556	47.3	5.1	17.1	30.2	28.0	19.2	7.0	107.9
3	0.072	0.038	0.303	0.067	0.581	48.2	5.0	19.6	28.6	28.6	19.6	7.3	103.2
4	0.097	0.042	0.328	0.051	0.573	57.4	6.0	19.5	37.9	34.0	23.3	8.3	104.8
5	0.080	0.042	0.312	0.072	0.573	57.4	6.0	17.9	39.5	34.0	23.3	8.3	104.8
6	0.097	0.042	0.337	0.046	0.577	58.1	6.0	20.2	37.9	34.5	23.6	8.5	104.0
7	0.076	0.038	0.312	0.063	0.569	48.5	5.1	16.7	31.8	28.8	19.7	7.4	105.5
8	0.093	0.038	0.320	0.046	0.560	47.9	5.1	16.6	31.2	28.4	19.5	7.2	107.1
9	0.093	0.038	0.316	0.046	0.564	47.6	5.1	18.1	29.5	28.2	19.4	7.1	106.3
10	0.105	0.042	0.328	0.038	0.573	56.6	5.9	22.5	34.1	33.6	23.0	8.0	104.8
11	0.101	0.042	0.316	0.046	0.560	55.8	6.0	19.9	35.9	33.1	22.7	7.8	107.1
12	0.084	0.038	0.307	0.059	0.573	47.6	5.0	18.4	29.2	28.2	19.4	7.1	104.8

Comparative analysis of example No. 19 from practice

Phase duration.

The QRS complexes are equal in their duration.

A difference in the phase duration between the PQ phases is 0.065s.

A difference in the phase duration between the QT phases is 0.044s.

Diagnoses.

Classical diagnosis option: myocardial hypertrophy of the right atrium & the right ventricle. Incomplete right bundle-branch block.

The phase analysis shows the following.

Of special note is a weak function of the lungs. The RHEO is like a breathing wave. This is an indication of a complicated pathology of the lungs. The wave P has a very great amplitude, and the phase PQ shows a significant dip. This is an indication of a significant pathology of the heart as a whole.

Recommended should be an appropriate in-hospital therapy.

10. Some special cases from practice

The CARDIOCODE heart analyzer application makes it possible to determine every diagnosis more precisely. The prerequisite is a certain experience to be accumulated by personnel in the application of this analyzer. This chapter describes some data obtained in monitoring patients in heavy pathological cases whereby each patient was monitored during more than one year period.

10.1. Phase analysis data obtained during pre-stroke & stroke periods and resuscitation & after-therapy periods

Patient XX20 was examined, and he did not present any heart area pain complaints. The patient evaluated by himself his general state of health as satisfactory.

When registering the patient's cardiosig-nals in lying position, their shapes were obtained as given in the Figure below (Fig. 110).

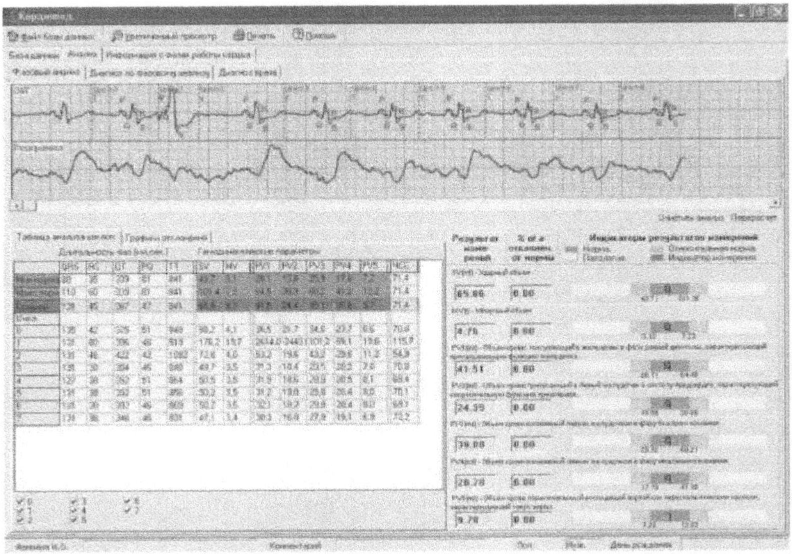

Fig. 110. ECG & RHEO of patient XX20 in lying position

The above figure shows that a bifurcation of the R-wave on the relevant ECG diagram indicates that there is a disturbance in conductivity of His' bundle. Extrasystole in cycle 1 is noted. In this case, the shape of the QRS complex is close to that featuring an acute development of infarctation. But the patient's hemodynamic parameters are within the norm. The relevant RHEO shows a relatively normal change in his arterial pressure (AP) in aorta. At the time of examination there were no patient's symptoms of a disease discovered, and the patient did not present any complaints.

The patient was recommended to be examined more closely considering at the same time the fact his last examination took place long ago.

According to the information by the patient, he visited every week a sauna where plentiful food was usually taken. He did not exercise any physical activity; his job was connected with mental activity.

The patient stated that four month ago a stroke occurred as he was outside in a street. He fainted and fell. The patient was brought to a hospital by the medical emergency service.

The patient was examined more carefully during resuscitation period, with an appropriate orthostatic test included.

Figure 111 shows the cardiosignals registered in the patient's lying position.

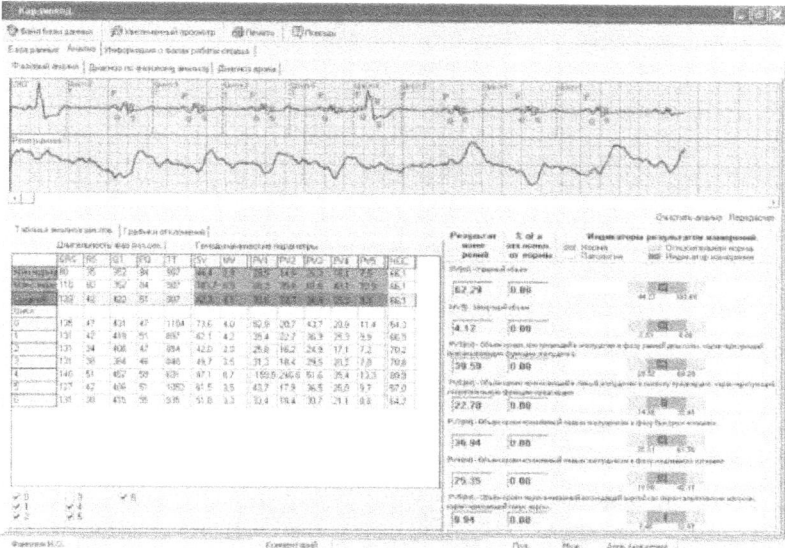

Fig. 111. ECG & RHEO of the patient during resuscitation in lying position, period: three days after stroke

It is evident that the amplitudes of the QRS complex are very small. Periodically they may have greater hypertrophied amplitude which is similar to that registered four months ago under the patient's ex-trasystole conditions. The RHEO shows that the AP changes detected are within the norm. It should be noted that the he-mody-namic parameters are within the norm, too.

397

The stroke resulted in a partial loss of movement functions by the right part of the body.

The shapes of the cardiosignals in sitting position show their remarkable changes (please, refer to Figure 112). There are no extrasystoles available. The ECG diagram shows waves of great amplitudes in the tension phase and in the aortic valve opening phase. There are the AP hemopulses of negative shape on the RHEO diagram in cycles 1 & 3 available, and they indicate that there is an ischemic stroke available, and this detection is supported by the diagnoses made by the doctors in charge. When evaluating the relevant RHEO, a reduction in venous blood flow can be discovered.

The patient was examined in lying position at an angle of 450 when cushions were placed under the patient's back. The ECG diagram produced under such conditions is given in Figure 113 below.

а)

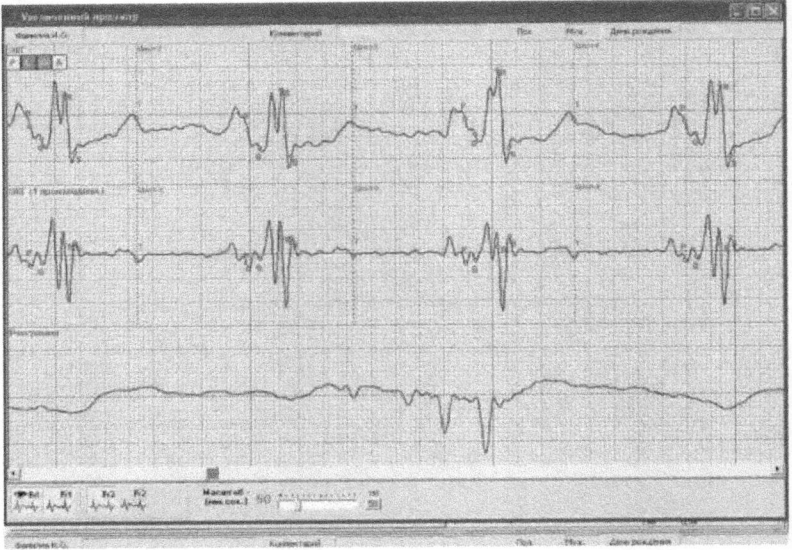

b)

Fig. 112. ECG & RHEO of patient XX20, period: 3 days after stroke, during re-suscitation. Figure a) is applicable to sitting position. Figure b) zooming version

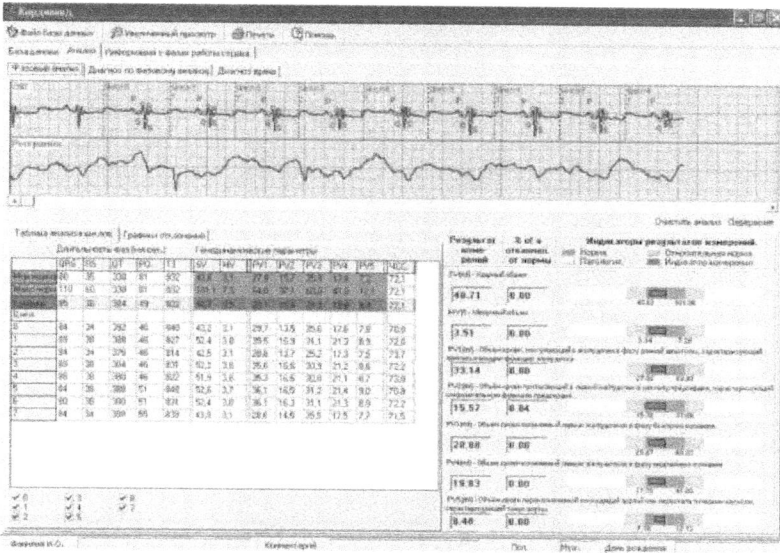

а)

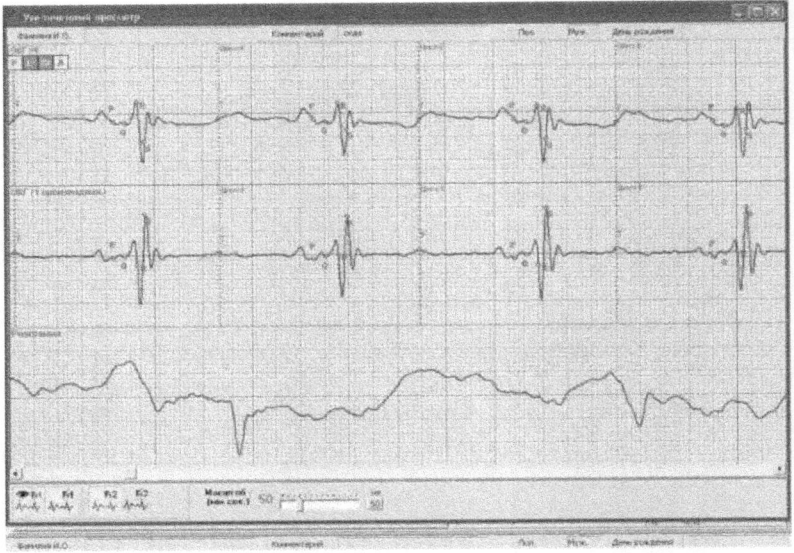

b)

Fig. 113. ECG & RHEO of patient XX20, period: three days after stroke, during resuscitation, lying position, at an angle of 45o to the horizontal

We think this shape of the cardiosignals is the most acceptable for an intensive therapy since there are fewer phases where the heart & blood vessel are loaded. The position pre-selected for the patient based on this analysis was recommended for him as the most favourable both for day and night periods.

Patient XX20 left hospital with partial loss of movement functions of the right part of the body. But he was able to go and do his job. At the same time the patient was recognized to be a disable person belonging to a certain invalid group.

Nine months later the same patient was examined by the analyzer CARDIOCODE once again (Figure 114 below).

First, it is evident that there is an extrasystole in cycle 2 and an instability of the T-wave in cycle 8 available which is responsible for an increase in the duration of cycle 9 (Figure 114a). A closer look shows that the above mentioned instability of the QRS-complex is of small value. The RHEO diagram registers that there is a very low elasticity of the aorta available whose indication is a flat apex on the RHEO, and there is a reduction in the venous blood flow detected whose indication is an actually horizontal line on the RHEO after its apex. The hemodynamic parameters are within the norm. The only parameter – PV5 – the pumping function of the aorta – shows that the aorta functions under a load exceeding the normal value by 3.37%. But this parameter remains within the norm, too.

The examination of the above patient was completed with an appropriate orthostatic test (Figure 114, b).

According to this Figure, when changing the position of the body to the vertical, there are many artifacts registered. Characteristically, on the RHEO, there are the AP pulses of great amplitudes available. The artifacts on the ECG are to be referred to an interference or disturbance.

Upon expiration of several tens of seconds there are neither interference nor disturbance registered (Figure 114, c). But much more often the extrasystoles can be detected.

403

The most distorted shape of the QRS complex is found in cycle 11. In this case the disturbance in the myocardium conductivity can be noted. But then we have to deal with an improved shape of the ECG diagram whose shape is completely restored (Figure 114, d), and upon the shape improvement, the registration of the car-diosignals is stopped. The shape of these signals is very close to an ideal one. The RHEO shows in this period also normal phase changes in the AP.

The orthostatic test demonstrates an existing problem of residual changes in the body which may affect generation of the cardio-signals. Therefore, the patient is recommended to avoid sudden changes in his body positions, but at the same time to exercise periodically physical activities under control.

a) in lying position

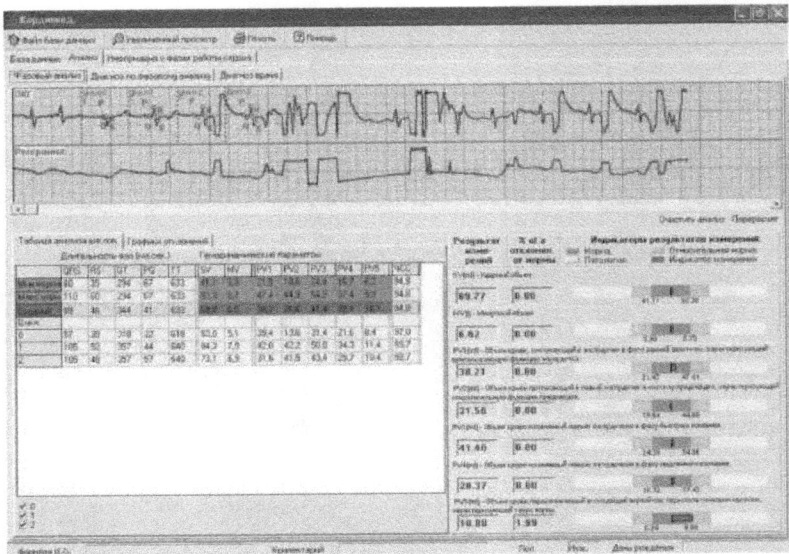

b) in sitting position, immediately upon leaving the horizontal position

c) in sitting position, upon expiration of one minute

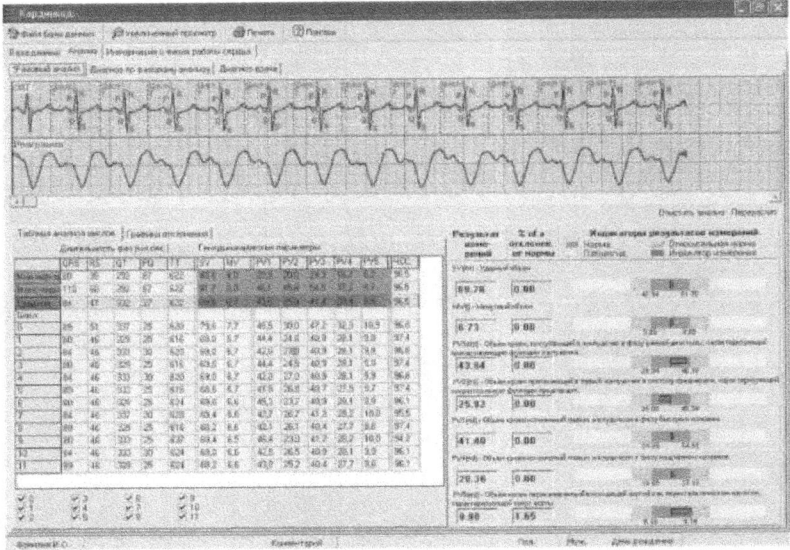

d) in sitting position, upon expiration of four minutes

Fig. 114. ECG & RHEO of patient XX20, period: nine months after stroke, both in lying & sitting positions

10.2. Another case of phase analysis data: stroke successfully avoided

A patient consulted us and presented his complaints as follows: severe pain in oc-cipit, eye-sight ability limited to a certain close distance before him, phobia against forward flexion of head and pains in the chest area. The patient detected all these symptoms by himself suddenly, when he was on the way to a shop. He had to give up shopping, went home and took his bed. The patient took some vasorelaxants. His arterial pressure was increased up to 150/90 mm wc. One day later he felt it getting better and visited his doctor. His doctor in charge prescribed him to take some medicines for several weeks.

But the examination with the heart analyzer CARDIOCODE produced the relevant patient's data as follows:

Figure 115, a shows that there are pulses of a short duration on the RHEO available characterizing changes in the AP in the ascending aorta. These pulses are called by us "hemodynamic pulses". The first pulse is produced at that instance when the aortic valve is fully opened, and the end of the pulse refers to that time when the aortic valve begins its closing (Figure 115, a). The second pulse is produced at that instance when the atrioventricular valve is fully closed, and the end of this pulse refers to that time when the aortic valve is fully opened. The third pulse is produced at that instance when the atrioventricu-lar valve starts its closing, and the end of this pulse refers to that time when the said valve is fully closed. The fourth and the fifth pulses correspond to the third pulse picture. The sixth pulse corresponds to the first one.
These features show that there is an interconnection between the production of the said pulses and the function of the heart valves. At the time of initiation of the 1st, 3rd, 5th and 6th pulses there

are fluctuations of the potential with greater amplitude on the ECG available.

Under close examination, other hemopulses are discovered, too. Figures 115b–d show their shapes.

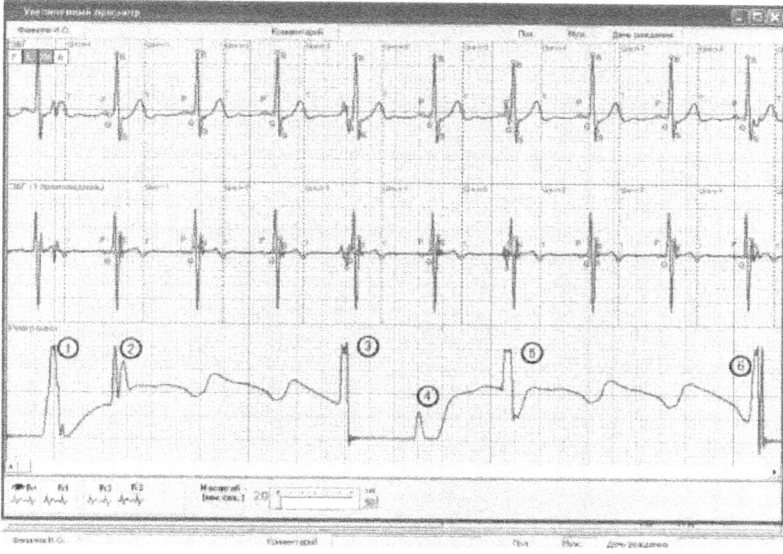

Fig. 115a. ECG & RHEO of patient XX21

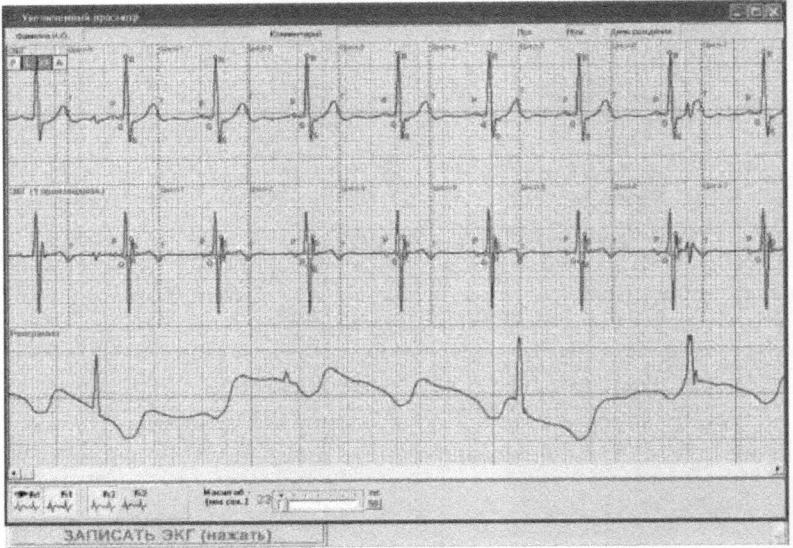

b)

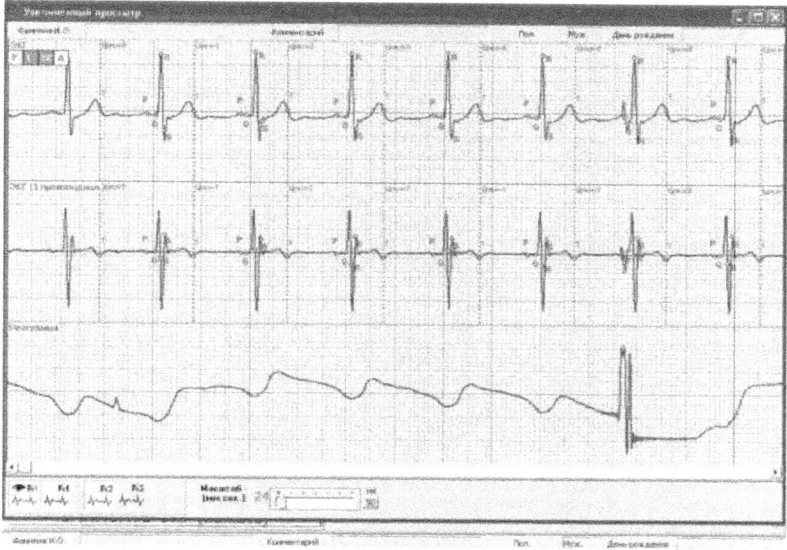

c)

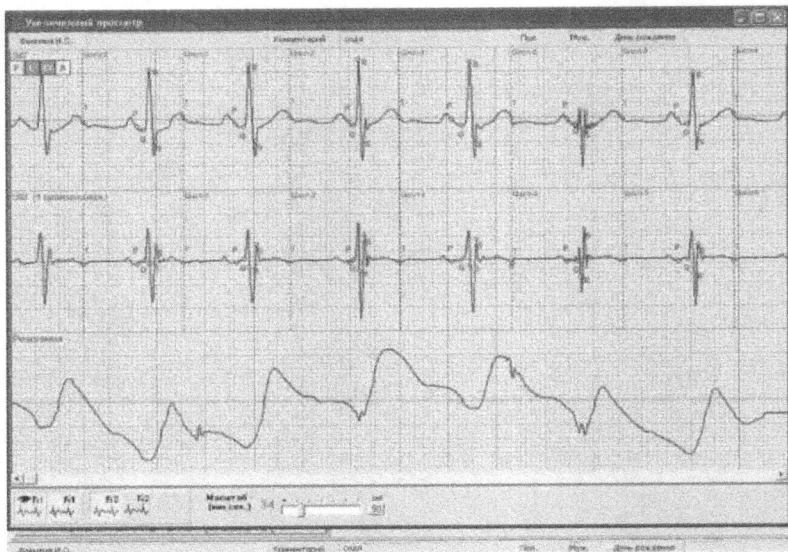

d)

Fig. 115 b,c,d. ECG & RHEO of patient XX21 upon close examination in details

414

Characteristically, the hemopulses produced in sitting position are of very small amplitude (Fig. 115, d). The patient states that his general state of health in the morning, immediately upon waking-up, is worse than that during his waking. Therefore, the patient is recommended to place the bed in such a way so that it is slightly inclined to minimize amplitudes of eventual hemopulses during sleeping.

The medicines prescribed by the patient's doctor in charge have not produced the desired effect so that the patient is prescribed acupuncture massage in pain areas in the chest zone. The patient is also recommended to take a medicine controlling metabolic processes referring to anti-oxidants, but with a moderate hypocholes-terolemic effect. Certain inhibitors of free radical processes in the body area stabilize the membrane structures of the blood vessel walls. One of these medicines is known in pharmacy under trade name "Mexicor".

Patient XX21 visited the doctor two weeks later again in order to make diagnostics. According to the patient's information, his general state of health was improved, but periodically he suffered from pains in the chest. The diagnostics demonstrated (Figure 116a, b) that there are no hemopulses in the horizontal position available. Only a minor increase in the AP is registered in this case.

One hemopulse is registered during the patient's orthostatic test in sitting position (Figure 116b). This corresponds to a change in the ECG diagram shape. But evaluating the changes in the hemodynamics within this cycle, it could be stated that the hemodynamic parameters remain within the norm.

Ten days later the patient visited his doctor again because of deterioration of his general state of health. In his opinion, it was associated with negative emotions because of his close friend's death. The data obtained in the examination show that the hemopulses occur, i.e., they are re-produced again (Figure 117a,b).

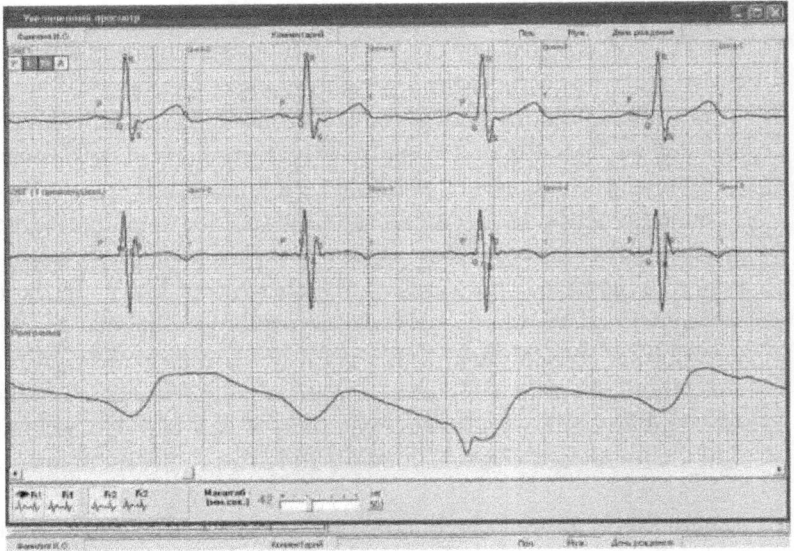

а)

416

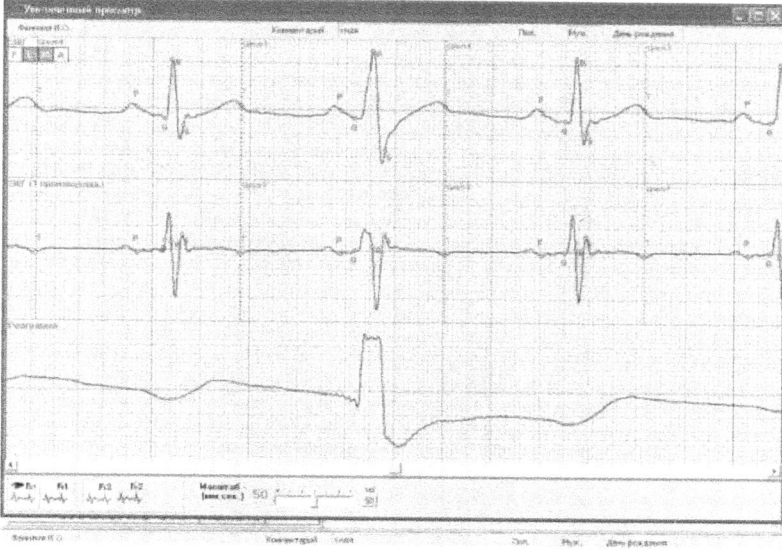

b)
Fig. 116 a,b. ECG & RHEO of patient XX21 under re-examination, two weeks later

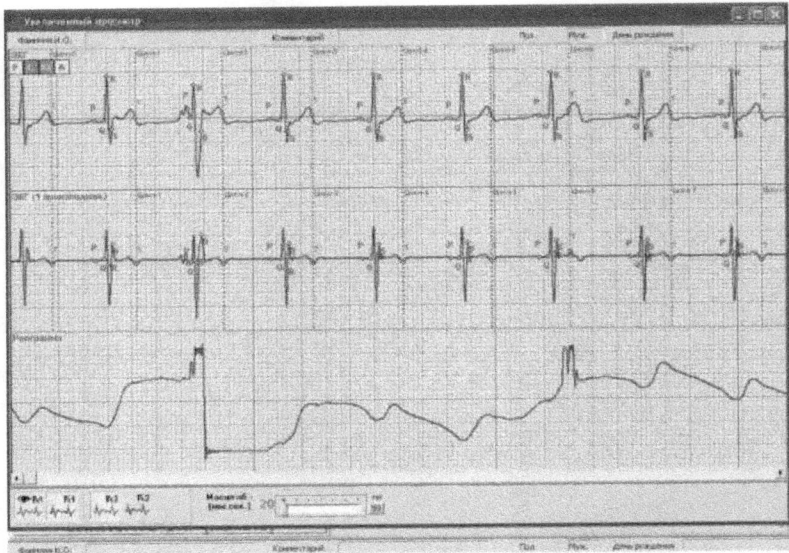

а)

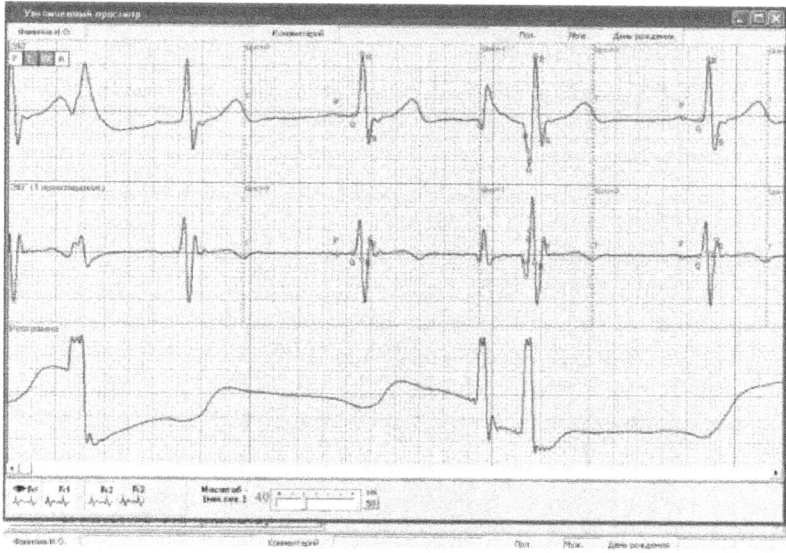

b)
Fig. 117 a,b. ECG & RHEO of patient XX21 in next diagnostics

The patient's symptoms were as follows: headaches, discomfort in cervical part of spinal cord, and an increase in the AP.

The patient was recommended to complete an acupuncture course and take herbal tea available at chemist's which can clean blood vessels.

The patient completed 13 acupuncture sessions with an interval of several days.

Six months later the patient was examined again (Fig. 118).

The health state of the patient was improved. The hemodynamic parameters reached their normal values again. The only noticeable fact was that the wave S amplitude on the ECG diagram decreased, and that indicated a diminished contraction function of the ventricles. The single he-mopulse was detected, but of apex type.

Two months later that patient was examined again. There were no symptoms presented which were registered previously, and that was an indication of a positive effect of the treatment, but the complaints about periodical headaches were presented. Upon the examination, there was detected that the main cause was the development of stenosis (Figure 119 a,b).

To treat stenosis, the patient did special breathing exercises with use of a special medical instrument – capnicator bought by him for this purpose (trade mark SAMOS-DRAV). This instrument should be used to reduce and normalize concentration of carbon dioxide in blood. The patient restarted drinking the herbal tea that was prescribed by his doctor earlier.

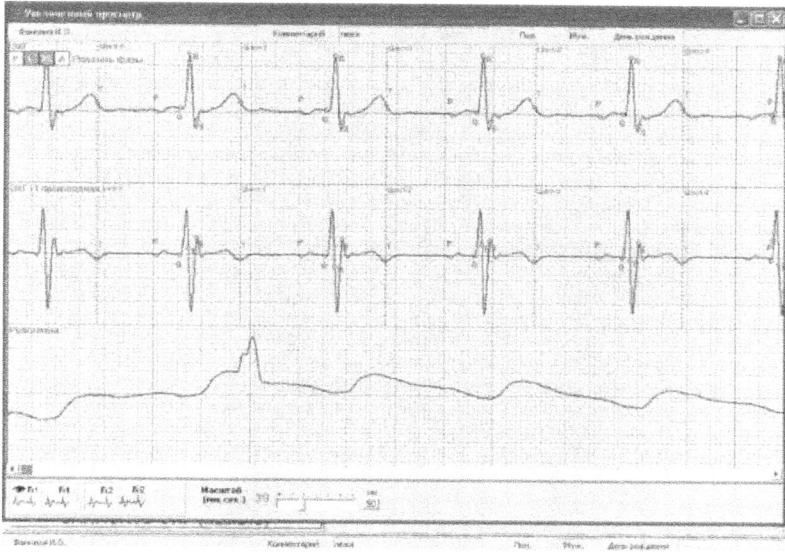

Fig. 118. ECG & RHEO of patient XX21 during the next examination

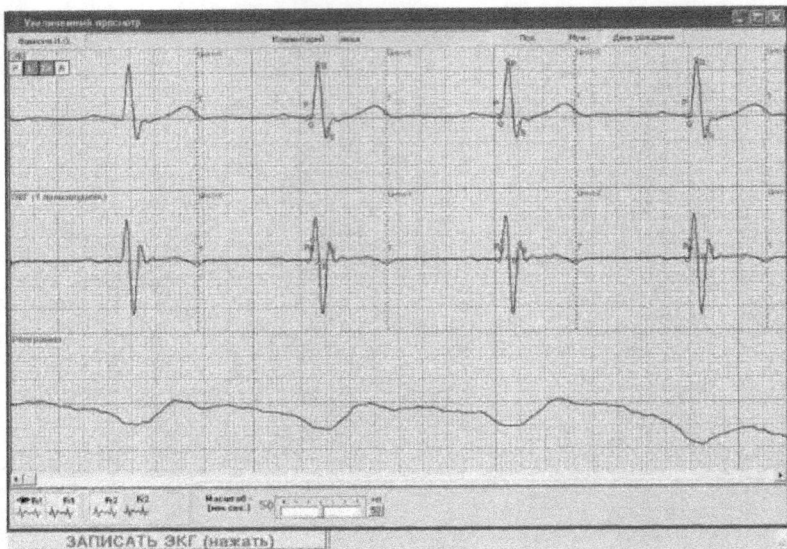

а)

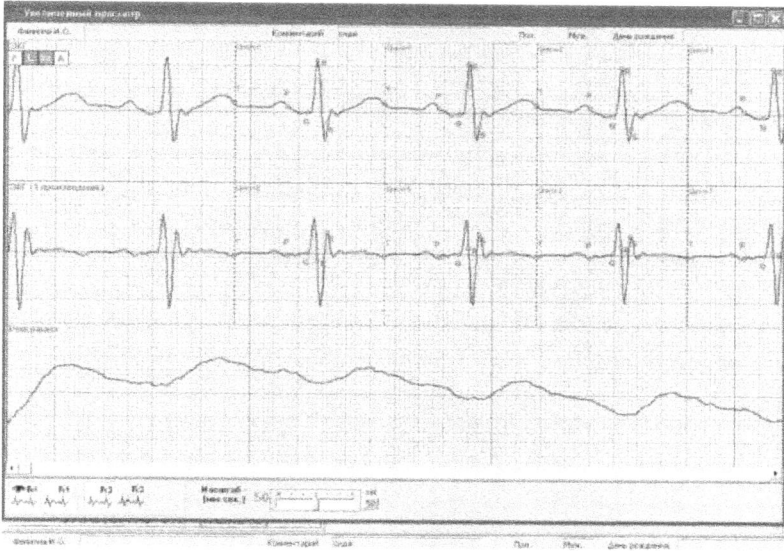

b)
Fig. 119 a,b. ECG & RHEO of patient XX21 indicating development of arterial
stenosis and stagnation in venous blood flow

Two months later the patient stated that no pain symptoms were available and that his general state of health was improved. The results of his reexamination were positive (Figure 120). Signs of stenosis were not found.

In summary, the description of this patient case should be completed by the information that the patient's job is associated with wood processing so that he is exposed to hazardous effects of painting materials and processed wood products. Considering this negative influence, it should be noted that the positive results have been achieved in his treatment owing to the doctor-patient dialog properly managed and controlled. In the above case used were well-known general means and treatment methods. The decisive factor of the successful treatment was support in supplying the most reliable data of the phase characteristics of cardiovascular system enabling to control the entire therapy course.

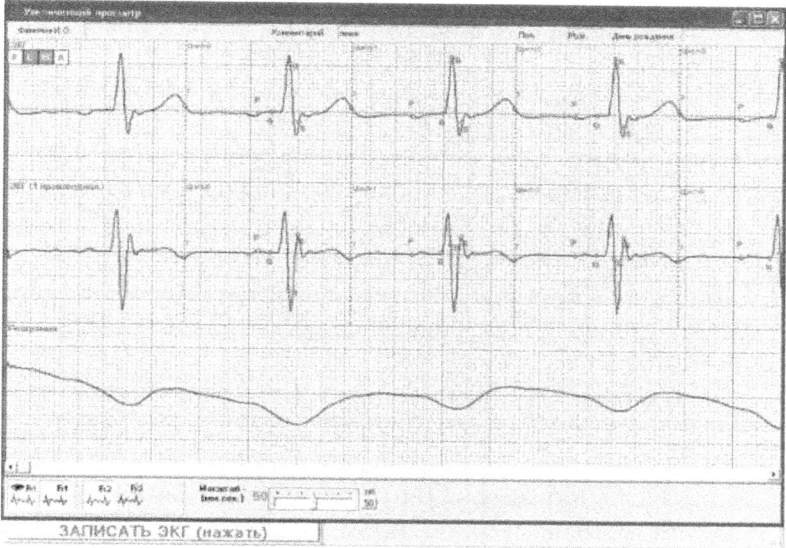

Fig. 120. ECG & RHEO of patient XX21 under latest examination

10.3. Data of phase analysis of patient with implanted heart pacemaker

Female patient XX21 contacted us and presented her complaints about her general health condition. During the consultation she informed us that she had an implanted heart pacemaker. An examination performed produced the following hemo-dynamic data:

This figure shows that the measured parameters of hemodynamics are extremely higher than their norm. All phase conditions are overloaded.

Let us analyze the salient features of the ECG diagram that is a product under influence of the pacemaker function. In all cases, a well-defined dip of the wave Q is an indication of the wave basis responsible for formation of the complex of the heart valves connected with ventricular septum. A decrease in the Q wave amplitude is the first indication of weakness of the entire basis of the valve system. In this case, the ECG diagram shows that the amplitude of the wave R is practically within the norm. But it seems that this wave is shifted and located below the isoline since the apex of this the apex of the wave wave is the same in its shape as P.

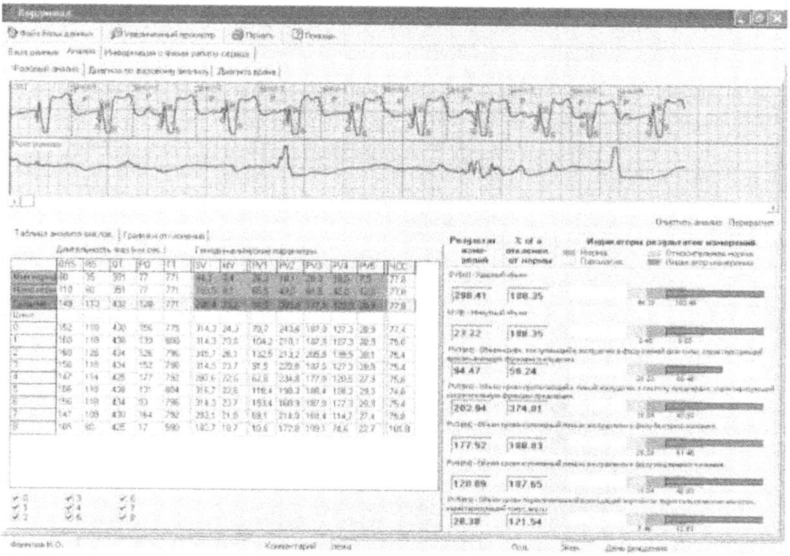

Fig. 121. ECG & RHEO of patient with implanted heart pacemaker

Figure 122 shows more details covering this problem.

Of interest is the S complex in this case. Its shape shows that there is a heavy pathology available. A bifurcation of the S wave is an indication of a disturbance in conductivity in the ventricles. After point S there is a diagram rise available exceeding the wave R. Considering this fact together with the above mentioned bifurcation, it should be noted that there is a heavy heart pathology detected. This can be supported by the relevant hemodynamic data (s. Figure 121). Only one parameter – PV1 – volume of blood entering the ventricle in the early diastole phase – shows the greatest deviation of 56% thus exceeding considerably the norm.

But despite this fact the heart is heavily overloaded in the other phases. These parameters can be interpreted by the heart self-control or self-regulation. Parameter PV1 indicating the degree of filling

427

the heart with blood is highly variable and is associated with the generation of the wave P. Figure 121 shows that it is not generated in every cycle. Under such conditions, the required volume of blood in the ventricles is provided only due to an extension of the ventricles. But this volume is desicive for heart overloading. The wave P occurs in those cycles when parameter PV2 characterizing the contraction function of the atrium reaches its relatively low level. Next cycles show that the wave P degenerates. This is a cyclic process that might be followed in Figures 121 and 122. This process is well defined through hemody-namic parameters PV1 and PV2 indicated in the relevant table in Figure 121.

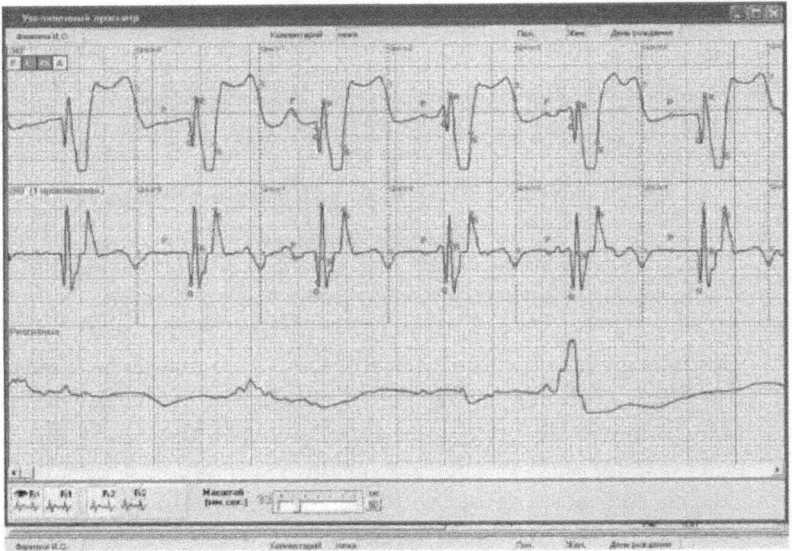

Fig. 122. Salient features of ECG diagram of patient with implanted heart pacemaker

The changes in the AP of the aortic blood registered on the RHEO in lying position are the minimum (Figure 122). Some he-mopulses occur in this case.

But during the orthostatic test we have to do with significant changes in the general picture (Figures 123 and 124). The development of the changes in the AP (Figure 123) registered in sitting position was normalized. Attention should be paid to an increase in value PV1. Under these conditions, the wave P decreases its amplitude and degenerates from cycle to cycle. This process is definitely registered.

The aortic valve in sitting position undertakes its functions normally. An increase in the AP both in the rapid ejection and the slow ejection phases is within the norm. The venous blood flow is normal. General health condition of the patient in her vertical position is improved.

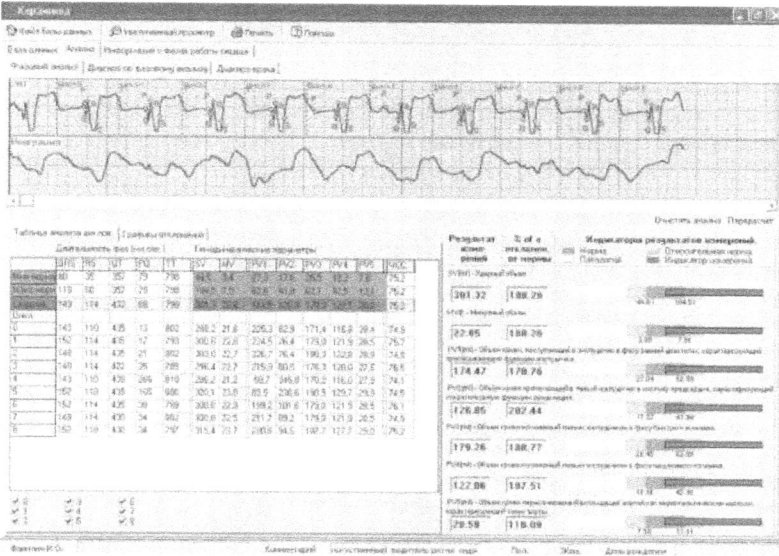

Fig. 123. ECG & RHEO in sitting position

The above analysis could be added by the data received from the doctors who had implanted the pacemaker and who are responsible for her monitoring. They informed that the only criterion of the normal operation of the pacemaker is to monitor the preset threshold value of the pacemaker generator. Other parameters of the cardiovascular system functioning should not be monitored. This is not an obstacle for persons with implanted pacemakers when some of them complete their 15 pacemaker life years.

It is beyond doubt that the application of the phase analysis makes it possible to get more answers to questions which may occur during the development and implantation of the pacemakers.

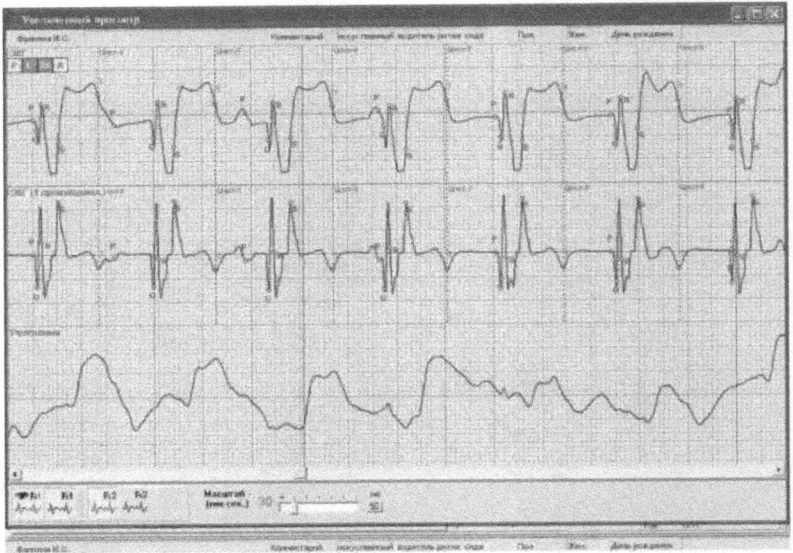

Fig. 124. ECG & RHEO in sitting position, zooming version

10.4. Data of phase analysis of patients with implanted artificial aortic valve and artificial ascending aorta

Patient XX22 consulted his doctor and was examined in order to define his hemo-dynamic parameters more exactly after he had been implanted an artificial aortic valve one year ago. At the same time a portion of his aorta had been replaced with an artificial aorta. The operation had been carried out due to heavy cordial hypertrophy that was not acceptable for his normal life.

Figures 125 and 126 show the results of the examination of this patient in lying position. It is evident that the wave R is not available at all on the ECG in every cycle. This is an indication of complete failure of the contraction function of the ventricular septum.

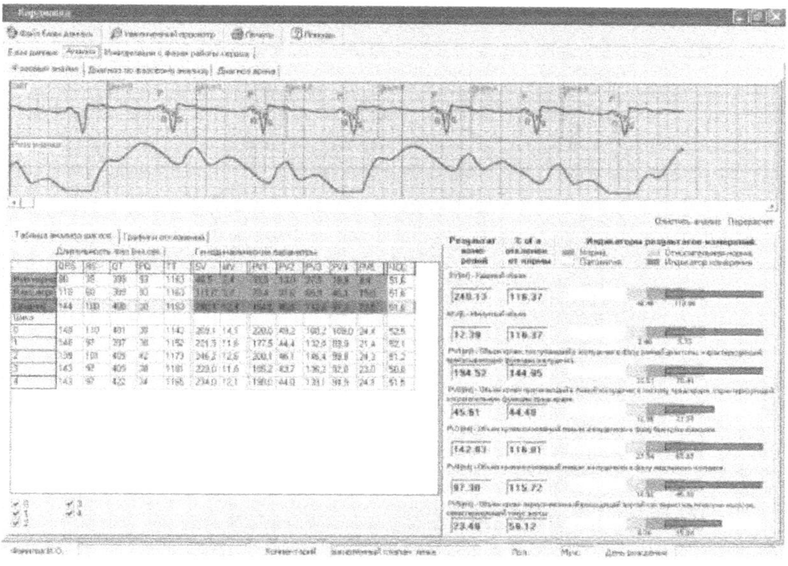

Fig. 125. ECG & RHEO of patient with implanted artificial aortic valve & artificial ascending aorta in lying position

431

The first task is to check how the artificial aortic valve performs its functions. At the time of registering the point S, by the beginning of the tension phase of the valve, there is no pressure increase in the aorta detected. Then, the valve opening occurs at the same time as the AP in the aorta starts its increasing. This is in conformity with an absolute norm of the valve functioning. The pressure development in the phases of rapid and slow ejection is O.K.

Special attention should be paid to the wave T whose amplitude is the minimum. Considering 56% deviation of parameter PV5 from the norm characterizing the pumping function of the aorta, it becomes evident that the aorta is overloaded. Under these conditions, after the wave T, a considerable decrease in the AP on the RHEO is detectable, and that is an indication of aorta flabbiness. In this case it should be referred to the artificial portion of the aorta.

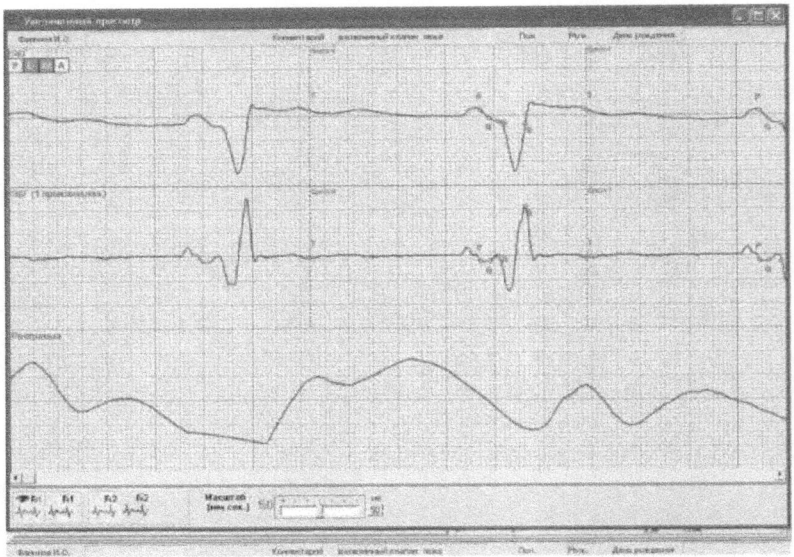

Fig. 126. ECG & RHEO of patient with implanted artificial aortic valve & artificial ascending aorta, zooming version

The orthostatic test shows that the effect of the aorta flabbiness in vertical position is not available (Figures 127 & 128). The aortic valve functions well, too. An increase in the AP starts strictly at the time of the valve opening.

All hemodynamic parameters significantly exceed their normal values. Of importance should be an increased width of the S wave. But on time the deviations from their normal values diminish (Figure 129).

We are not able to comment the identical changes on the ECG occurring in every third cycle (Figure 130).

When making the prescribed diagnostics, the patient presented his complaints about a minor disorder in his health conditions. In our opinion, the cause that led to hypertrophy of the ventricles was not eliminated. This followed from the orthostatic test showing a change in the S wave width.

In spite of the complicated operation associated with implantation of the artificial aortic valve & a portion of the artificial ascending aorta, it can be stated with certainty: the operation successfully has improved the general state of the patient, and this can be supported by the relevant data of the phase analysis of the heart cycles.

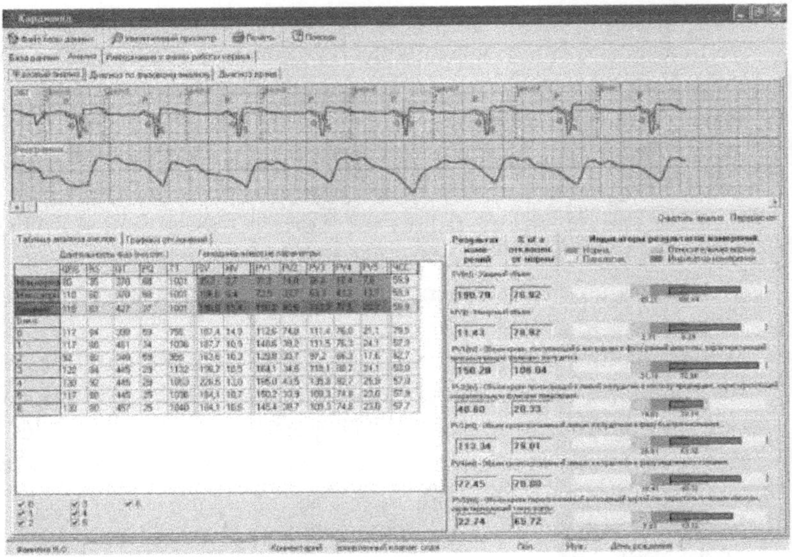

Fig. 127. ECG & RHEO of patient with implanted artificial aortic valve & artificial ascending aorta in orthostatic test, vertical position

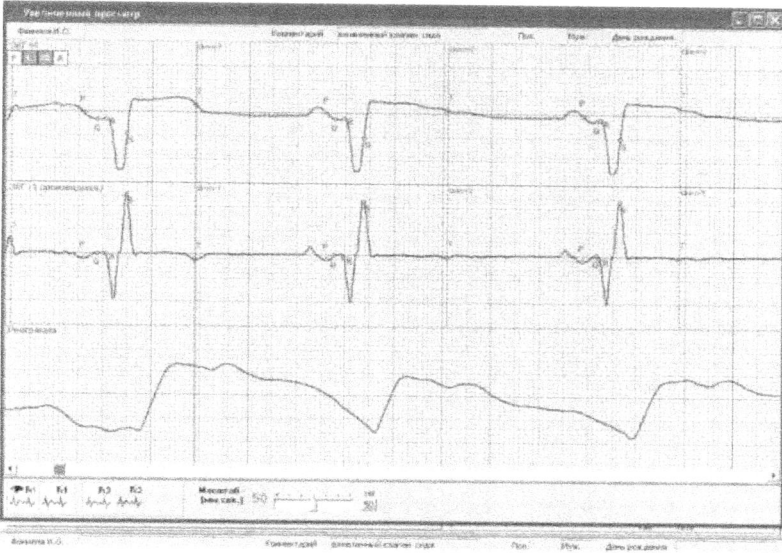

Fig. 128. ECG & RHEO of patient with implanted artificial aortic valve & artificial ascending aorta in orthostatic test, vertical position, zooming version

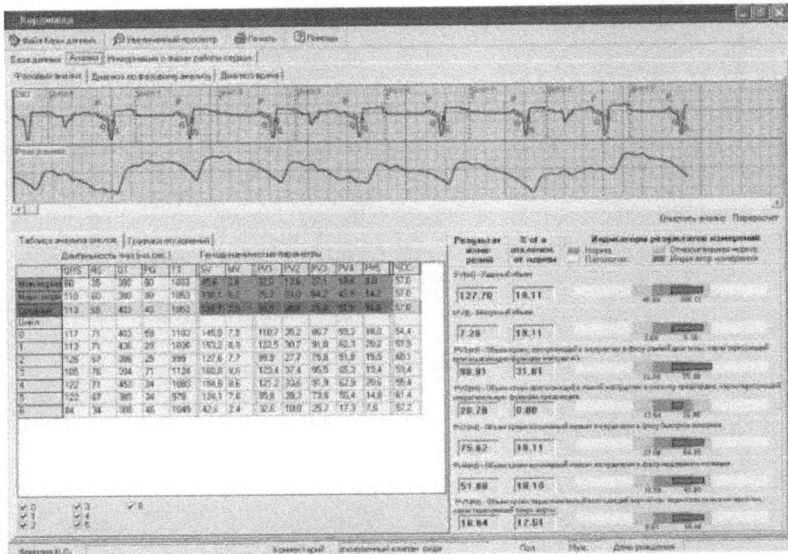

Fig. 129. Deviation of hemodynamic parameters diminishes on time

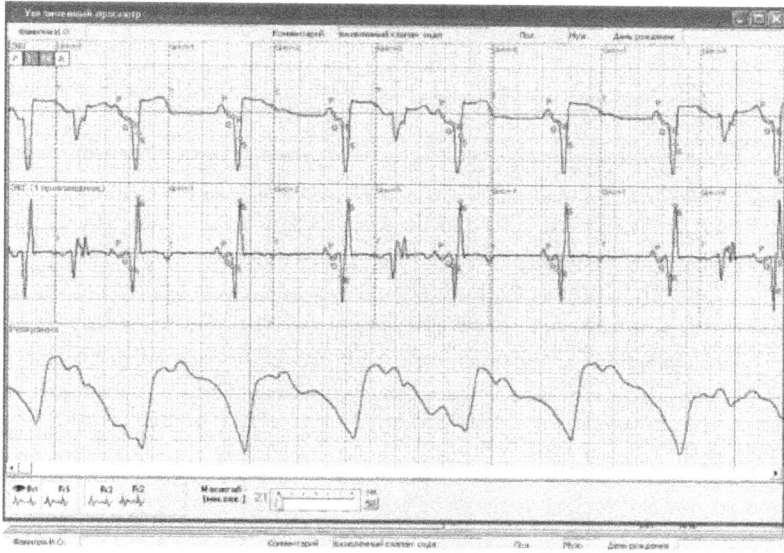

Fig. 130. Fluctuations caused by unknown factor occur on the ECG in every third cycle

11. Phytotherapy of heart diseases

Prior to describe the most efficient phyto-therapy means used for treatment of cardiovascular system diseases that are proven in practice, it should be noted that the therapy shall cover a complex of treatment measures. The greatest efficiency can be achieved when a combination of several treatment measures is applied at the same time: acupuncture, phytotherpay, synthetic medicaments, controllable physical activities supported by well-balanced nutrition. This chapter offers well-known approved phytotherapy formulae.

11.1. Case of cardiac abnormalities: weak function of heart muscle, manifestations of cardiac insufficiency

1). Fruits of apricot tree (Prunus armeniaca) are widely used in Tibet Traditional Therapy of anemia and other cardiovascular diseases since the apricot fruits contain potassium salts removing water from the body.

2). Fill a 0.5 l bottle with Adonis herb finely cut and add 56% vodka. Hold it for 12 days at a warm dark place, shaking and turning over the bottle every day. Take 8 drops three times a day one hour before your meal. This phytotherapy medicine may be used for a long period of time.

3). Pound 0.5 kg ripe fruits of hawthorn (Crataegus) in a mortar with wooden pounder, add 0.5 glass of water, heat up to 400C and extract juice in a juice extractor. Take 1 table spoonful 3 times a day before your meal. Hawthorn has a positive effect on the heart; it is very important for aged persons since it prevents overstressing and wearing of the heart muscle.

4). Put one table spoonful of dry hawthorn fruits (Crataegus) into one glass of boiling water, hold it for 2 hours at a warm place (a thermos may be used for this purpose), then filter it. Take 1–2 table spoonfuls 3–4 times a day before your meal. Hawthorn-based medicines tonicize the heart muscle, enhance the coronary blood flow, eliminate arrhythmia and tachycardia, diminish excitability of the central nervous system and decrease blood pressure.

5). Put one table spoonful of valerian (Vale-riana) into 1 glass of boiled water cooled down to a room temperature, hold it in a closed vessel for 8–12 hours, and then filter it. Take 1 table spoonful 3–4 times a day.

6). Put one table spoonful of Uliginosi sprouts finely cut into 1 glass of boiling water, boil the mixture for 10 minutes slowly, and then cool down and filter it. Take 1 table spoonful 3 times a day in therapy of heart diseases.

7). Put erysimum herb (Erysimum) into 200 ml water. Take 1 tea spoonful 3 times a day as tranquilizer in therapy of heart diseases. In the Ancient Rome (under Plin-ium, Ist c. A.D.) the erysimum was well known and used as a good heart medicine.

8). Wild strawberry plants (Fragaria vesca L.) should be harvested with their roots in blossom phase and dried in shadow in ventilated rooms. 3–4 wild strawberry plants should be put into a teapot to make an infusion similar to tea; the infusion may be taken with sugar, and this therapy may be used for a month. This phytotherapy is the most efficient in cases of inorganic flaws of the heart.

9). Put one glass of snowball fruits (Vi-bumum opulus L.) into 1 l hot water, boil for 8–10 minutes, filter it and add 3 table spoonfuls of honey. Take 0.5 glass of the decoction 3–4 times a day. This therapy is of great efficiency when treating heart diseases and hypertension.

10). Put 1 tea spoonful of inflorescences of lily of the valley (Convallaria majalis L.) into 1 glass of boiling water and hold it for 30 minutes. Take 1 table spoonful 2–3 times a day. Lily of the valley is used in therapy both of acute & chronic cardiac insufficiency, cardiosclerosis and for enhancement of heart contractions.

11). Fill a glass vessel with a narrow neck with fresh inflorescences of lily of the valley (Convallaria majalis L.) (without stems) to 3/4, add 90% alcohol, close and hold it for 2 weeks, and then filter it. The tincture shall be yellow, clear and bitter to taste; it should slightly smell of lily of the valley. Take 10–15 drops 3 times a day as a medicine controlling heart activity, normalizing abnormal heartbeats and eliminating nervous excitation. The tincture based on lily of the valley shall be classified to a group of the most potent preparations, and therefore there are maximum prescribed norms fixed: a single dose shall be under 30 drops and the total day dose shall not exceed 90 drops.

12). Put one tea spoonful of dried or powdered peppermint leaves (Mentha piperita L.) into 1 glass of boiling water, put the lid on the vessel with the infusion, hold it for 20 minutes and then filter it. Take the infusion fasting, 30 minutes before your meal. Take it every day for several months. It should be used in case of a weaken heart – when heart activity is intermittent or tardy, etc.

13). Put oat seeds into water at a ratio 1:10, hold the infusion for 24 hours and then filter it. Take 0.5 glass 2–3 times a day before your meals. Use as diuretic preparation and as a medicine regulating metabolism processes in the heart muscle and nervous tissue.

14). Put four table spoonfuls of raw moth-erwort (Leonurus car-
diaca) (plant top part in blossom up to 40 cm long, under 4 mm
thick) into hot water, put the lid on the vessel with the infusion and
heat up in boiling water bath, stirring often the herbal tea, for 15
minutes, and then cool it down for 45 minutes up to a room temperature, and press the herbal cake. Then boiled water should be
added to the prepared infusion to obtain the full glass. Take 1/3 of
glass 2 times a day 1 hour before your meal. The infusion should be
kept at a cool place no longer than for 48 hours.

Another recommendation is to prepare the motherwort-based tincture using 70% alcohol, with a ratio 1:5. Take 30–50 drops 3–4
times a day.

15). Beetroot water-based infusion should be recommended in therapy of heart flaw diseases.

16). Put two tea spoonfuls of Lamiaceae into 1 glass of cooled boiled
water, hold for 4 hours and then filter it. Take 1/3 of glass 3 times a
day before your meal to regulate properly the heart function. No use
in case of hypertension!

17). Cottage cheese as a high-quality diet product should be taken
at a regular basis every day in the evening.

18). Take flower honey.

11.2. Herbal medicines recommended against heartbeats attacks

1). Heat up 1/4 l water close to the boiling temperature and reduce heat input, put 1 table spoonful of Adonis herb. Boil at a low heat for less than 3 minutes. Then put the lid on the pan, keep it at a warm place for 20 minutes to obtain an infusion, and then filter it. Take 1 table spoonful 3 times a day. Heartbeats attacks could be eliminated after several days of this therapy.

2). Put 1–2 tea spoonfuls of cornflower inflorescences into 1 glass of boiling water, hold for 1 hour and then filter it. Take 1/4 of glass 3 times a day 10–15 minutes before meal. Use in cases of heartbeats attacks.

3). Put one table spoonful of inula (Inu-la) into 2 glasses of water, boil slowly for 15 minutes. Take 2 table spoonfuls every hour for one day in case of heartbeats attacks. DO NOT USE inula under kidney diseases and pregnancy!

4). Put 1 table spoonful of Scutellaria bai-calensis Georgi roots into 100 ml alcohol. Take 25 drops 3 times a day thirty minutes before your meal. Use in therapy of myocarditis (inflammation of heart muscle) and frequent heartbeats.

5). Mix extracted juice of thousand-leaf plants (Achilea millefolium L.) with ex-tracte d ju ic e of Ruta grave olen s L . i n e qu a l parts. Take 24 drops per glass of vine or vodka. Take it 2 times a day in case of severe heartbeats.

11.3. Herbal medicines used to eliminate edema caused by cardiac insufficiency

1). Since ancient times flesh (pulp) of water melon is considered to be an excellent diuretic medicine used against edema caused by cardiovascular diseases and diseases of kidneys. A decoction prepared from fresh water melon shells shows very good diuretic properties, too.

2). Take 40 g dried roots of Levisticum of-ficinale Koch, boil them for 7–8 minutes in 1 l water and hold the infusion for 20 minutes at a warm place. Take the herbal medicine 4 times a day, only fresh made. It should be used if nephrocardiac edema is available in order to improve heart functioning.

3). Cucumbers being considered to be an effective diuretic medicine may be recommended for diets of persons suffering from hydrops and edema associated with cardiac diseases.

4). Parsley seeds should be pre-powdered. Four tea spoonfuls of the prepared seeds should be put into 1 glass of boiling water, boiled for 15 minutes, cooled down and filtered. Take 1 table spoonful of the infusion 4–6 times a day.

5). 1 tea spoonful of parsley seeds should be infused in 1 glass of water for 8 hours. Take 1/4 of the prepared infusion 4 times a day.

6). Use juice extracted from horsetail (Equisetum arvense L.). Juice should be produced only from the plants harvested early in the morning when dew occurs. The plants to be processed should be washed with running water, then the rest of water should be removed, and the plants should be put into boiling water for a short time. Then the clean raw plants should be minced, and the juice should be extracted and boiled for 2–3 minutes. The prepared juice should be kept in refrigerator. Take 1 table spoonful of the juice 3–4 times a day.

11.4. Traditional phytotherapy medicines used for relieving pains in heart area under angina pectoris

1). The use of cherry fruits reduces frequency of attacks and soothes pains in the heart area.

2). Take 10 fruits of juiniper (Juniperus L.), chew thoroughly and swallow them. Use as a pain-relief – pains should be relieved in 15–20 minutes.

3). Take three tea spoonfuls of asparagus roots with sprouts and leaves (Asparagus officinalis L.) finely cut and put them into 1 glass of boiling water, isolate to prevent heat loss and hold for 2 hours, and then filter the infusion. Take 1 table spoonful every 2 hours.

4). Thousand-leaf tincture (Achilea mille-folium L.) (1.5 g raw herb should be put into 100 ml vodka, holding time should be 7 days). Take 20 drops 3 times a day before meal.

11.5. Recommendations: how to prevent angina pectoris attacks

1). One tea spoonful of red clever flowers should be put in 1 glass of hot water; the herbal tea should be boiled for 5 minutes and then filtered. Take 1 table spoonful of the infusion 4–5 times a day.

2). Take 1 table spoonful of roots of Sene-cio platyphyllus Bieb DC. and put them into 200 ml boiling water. Take 40 drops once a day. It should be used as a pain relieving medicine in case of severe pains associated with an angina pectoris attack.

3). Tincture based on roots of Senecio platyphyllus Bieb DC. – 1.5 table spoonful of roots per 100 ml of 70% alcohol. Take 30–40 drops once a day, under uninterrupted pains it should be taken three times a day.

4). Peel 300 g garlic, put it into a half-litre bottle and add alcohol. Hold it for 3 weeks. Take 20 drops thinned in 0.5 glass of milk every day. DO NOT USE this herbal medicine under kidney diseases!

11.6. Herbal teas in therapy of arrhythmia

1). Juice extracted from fresh thousand-leaf plants (Achilea mille-folium L.): take 20–30 drops thinned in 1/4 glass of water or dry grape vine.

2). Decoction prepared from wild strawberry leaves and berries (Fragaria vesca L.) (2 table spoonfuls of dried raw materials finely cut to be put into 200 ml boiling water). Boil the tea for 10 minutes slowly, hold it at a warm place, then filter it and take 1 table spoonful.

3). Infusion prepared from pansy (Viola tricolor L.) (2 table spoonfuls of dried plants finely cut per glass of boiling water), holding time should be 1.5 hour. Take 0.5 glass of the infusion 2 times a day.

11.7. Phytotherapy against heartbeats accompanied by edema

1). Decoction of French bean pods (1.5 table spoonfuls of dried pre-material finely cut put into 300 ml boiling water). Boil the infusion for 30 minutes, filter it and take 1/3 glass of the decoction 3 times a day 30 minutes before your meal.

11.8. Under heartbeats induced by nervous system

1). Decoction of roots incl. rootstock of valerian (Valeriana) (2 tea spoonfuls put into 200 ml boiling water). Take 1 table spoonful 3–4 times a day.

2). Lavender flower tincture (Lavandula officinalis) (3 tea spoonfuls should be put into 400 ml boiling water), the infusion holding time should be 10 minutes; upon filtering, take a single dose during the day.

11.9. Under heartbeats accompanied by insomnia

1). Mixed herbal tea components:
1.5 table spoonfuls of roots incl. root-stock of valerian (Valeriana);
1.5 table spoonful of Melissa offici-nalis L.;
2 table spoonful of thousand-leaf flowers (Achilea millefolium L.).
Should be put into 200 ml boiling water and then filtered. Take 1/3 of glass of the tea 3 times a day.

2). Mixed herbal tea components:
2 table spoonfuls of horsetail (Equise-tum arvense L.);
3 table spoonfuls of knotgrass (Poly-gonum aviculare);
5 table spoonfuls of redhaw hawthorn (Crataegus sanguinea).
Should be put into 200 ml boiling water, infused for 10 minutes and filtered. Take 1/3 of glass of the infusion 3 times a day.

11.10. Under tachycardia

1) Mixed herbal tea:
2 table spoonfuls of horsetail (Equise-tum arvense L.);
3 table spoonfuls of knotgrass (Poly-gonum aviculare);
5 table spoonfuls of redhaw hawthorn (Crataegus sanguinea).
The infusion (1 table spoonful of the mixture should be put into 200 ml boiling water) should be kept for a night in thermos and then filtered. Take 1/3 glass of the infusion 3–4 times a day.

11.11. Under post-infarction conditions

1). Fresh-made carrot juice should be taken permanently 4–5 times a day. Honey or sugar should be added to taste.

11.12. Under microinfarctation

1). Sirup prepared from fresh spruce fir buds (Picea Excelsa Link) (put fresh spruce fir buds into a glass vessel layer by layer, adding sugar between the layers, hold for 3–4 weeks). Take 1 tea spoonful of the prepared sirup 3 times a day.

11.13. Under coronary insufficiency (CHD)

1). Mix 250 g peeled garlic with 350 of liquid honey and hold for 7 days. Take 1 table spoonful 3 times a day before meal, a therapy period should be 1–2 months.

2). Stachys betoniciflora tincture (3 tea spoonfuls of raw material should be added to 200 ml vodka) should be held for 7 days and then filtered. Take 15–20 drops 3 times a day.

11.14. Under heart arrhythmia

1). Prepare the mixed herbal tea as follows: 2 table spoonfuls of roots of valerian (Valeriana) + 2 table spoonfuls of Angelica archangelica L. + 1.5 table spoonfuls of Menyanthes trifoliata L. + 1.5 table spoonfuls of peppermint (Mentha piperita L.).
Take 2 table spoonfuls of the mixture finely pre-cut and put them into 0.5 l boiling water, boil the mixed tea slowly, hold for 1 hour and then filter it. Take 1.3 of glass 3 times a day after meal. Take at the same time 1 tea spoonful of bee cell pollen together with honey in equal parts, 3 times a day.

11.15. Under extrasystole accompanied by hypotension

1). Prepare the mixed herbal tea as follows: 1.5 table spoonfuls of dog rose (Rosa canina L.) + 1 table spoonful of inflorescences of Antennaria dioica L. + 1 table spoonful of plantain leaves (Plantago) + 1 table spoonful of white birch leaves (Betula alba) + 1 table spoonful of peppermint (Mentha piperita L.) + 1 table spoonful of knotgrass (Polygonum aviculare) + 1 spoonful of Sedum L. + 1 table spoonful of Polemonium Caeruleum L. + 0.2 table spoonfuls of lily of the valley herb (Convallaria majalis L.). Take 50 g of the mixture finely pre-cut, put it into boiling water in thermos and hold it for 2 hours. Filter the herbal tea and take 1.3 of the glass 3 times a day. Take at the same time 1 tea spoonful of bee cell pollen together with 1 tea spoonful of honey 3 times a day.

11.16. Persons suffering from coronary heart disease (CHD) with normal arterial pressure

1). Prepare the mixed herbal tea as follows:
• 2 table spoonfuls of redhaw hawthorn fruits (Crataegus sanguinea);
• 1.5 table spoonful of peppermint leaves (Mentha piperita L.);
• 1 table spoonfuls of calendula (Calendula officinalis L.);
• 1 table spoonful of saga (Salvia L.);
• 1 table spoonful of veronica (Veronica gen.);
• a half of the table spoonful of dry laminaria;
• a half of the table spoonful of Meliotus officinalis Lam;
• a half of the table spoonful of young branches with leaves of white mistletoe (Viscum album);
• a half of the table spoonful of sagebrush herb (Artemisia vulgaris);
• a half of the table spoonful of wild strawberry (Fragaria vesca L.).
Take 2 table spoonfuls of the dry mixture finely pre-cut, put them into 0.5 l boiling water and boil slowly for 10 minutes. Hold the infusion for 1–2 hours, and then filter the herbal tea. Take 0.5 of the glass of this tea 3 times a day. At the same time, 1 tea spoonful of bee cell pollen & honey at a ratio 1:1 should be taken 2–3 times a day. A therapy period should be 4–6 weeks. Then, the therapy course should be interrupted for 2 weeks and resumed.

11.17. General recommendations for diet management under any cardiac diseases

1). Take a middle-size black radish. Cut off the top that is to be used later as a sealing cover. Make in the radish a hole sized like a pigeon egg. Put 2–3 table spoonfuls into the hole and then seal it using the pre-cut radish top like a lid. Place the prepared radish into a suitable pot and keep it in refrigerator. Watch every day if the juice is properly released. Take 1 table spoonful of this juice once a day in the morning before meal.

2). Chew fresh wild rose flower petals (Rosa Cinnamomea L.), take jam made of them.

3). Decoction of oak bark (Cortex quer-cus) prepared as follows: take 1 table spoonful of oak bark and put it into 200 ml boiling water. Take 1 table spoonful 2–3 times a day.

4). Put 1 table spoonful of dandelion flowers (Taraxum officinale Wigg) into 200 ml boiling water (in a glass or an enamel pan), hold the infusion for 20–30 minutes. Take 1/3 of glass 3 times a day before meal during 2 weeks.

5). Take 1 table spoonful wild chicory (Ci-chorium intybus L.) 2 times a day every two days.

6). Take 0.5 of glass of juice extracted from pome-granates 2 times a day.

7). Extract cabbage juice with an extractor. Take 0.5 of glass once a day before meal. The juice should be kept in a refrigerator no longer than for 24 hours.

8). Take flower honey in case of any cardiac disease.

9). Take 100 g cottage cheese under any heart and lever diseases every day.

10). Take lemon incl. peel and apricot fruits (dried apricot fruits).

11). Use for food baked unpeeled potatoes.

12). Put 0.5 of glass of dog rose fruits (Rosa canina) into 1 l boiling water in thermos and hold it for 12 hours. Take as the herbal tea 3–4 times a day.

13). Use for food as often as possible nuts, raisins and cheese in case of weakened heart function.

14). Nutmeg and milk are ideal for an improvement of heart functioning.

15). Tea made from dried quince fruits is useful against cardiac edema. In this case parsley, lemon, cucumbers and pumpkins are very useful.

16). Use garlic for food either pickled or fresh.

17). Use for food wild strawberry (Fragar-ia vesca L.) and take tea made from wild strawberry leaves.

18). Enjoy apricots. Drink one glass of apricot juice every day.

19). Lemons, figs, apples, pome-granates and pears are the most acceptable food against arrhythmia.

20). Use the tincture prepared from geranium (Geranium) with vodka (ratio 1:10) under heart pains. Take 20 drops of this tincture 3 times a day or in case of sudden acute pains.

Conclusion

The authors presented the relevant results of their researches obtained with application of the phase analysis of the heart cycles. The research activities have been carrying out by our R & D team for more than 25 years, but only 3 years ago the first positive results were obtained which make it possible to transfer from testing the validity of the theory to a widespread practice.

This book should be evaluated as an introduction into the theory & practice of the phase analysis of the heart cycles. The authors might express their hope that further work on this subject will lead to new discoveries, and that the materiel contained herein could be used as basis for the future advanced technologies and techniques.

But one thing is to be definitely stated: publishing of this book is the fist stage of formation of the new trend in cardiology: the development of the phase analysis of the heart cycles.

The analyzer CARDIOCODE developed by the R & D team within the framework of the CARDIOCODE project makes it easy to get an access to further research activities for every doctor concerned. At present, the researches in this field are continued, and the authors indent to publish another book – an interactive version of the study – which should contain more detailed descriptions of other specific pathologic cases.

Bibliography

1. A New Definition of Myocardial Infarction (a joint document of the Unified Committee of the European Cardiology Society and the American Cardilogists Collegium with respect to revision of the existing definition of myocardial infarction)// Journal of the American Cardiologists Collegium (JACC), Vol.36 No.3, 2000, September, P.959-969

2. Beryozov T.T. Application of Ferments in Medicine // Soros' Education Journal, Biology. 1996 No. 3, P.23-27

3. Caro C, Padley T., Shroter R., Sid W. Blood Circulation Mechanics, M.: "Mir», 1981. P.624

4. Scientific Discovery No.290 Law of Propagation of Arteric Pressure Waves in Blood Vessels in Areas of Local Increase in their Impedance / Alexeyev V.B., Zernov V.A., Matsyuk S.A., Rudenko M.Yu.. Application dd.05.07.2005, A-358; published 31.08.2005 [electronic resource:] Access: http://www.raen.ru/discoverv/288.shtml7in

5. Medical Electronic Equipment for Public Health Service, Transl. from English into Russian, M.Arditti, F.Waybell et al., Translation edited by R.I.Utyamysheva, M., "Radio I Svyaz", 1981

6. Shelagurov A.A. Propaedeutics of Internal Diseases. M., «Medicina», 1975

7. Andreyev L.B., Andreyeva N.B. Cinetocardiography. Rostov-on-Don, Published by the Rostov State University, 1971, 308 pages

8. Popetchitelyev Ye.P., Korenevsky N.A. Electrophysiological & Photometric Medical Equipment. M., «VysshayaShkola», 2002, P.208-212

9. Strutynsky A.V. Echocardiogram: Analysis & Interpretation. M., «Medpressinform», 2001, 208 pages

10. Digital Blood Pressure Measuring Instrument. Catalog: «Sensors produced by MOTOROLA». M., «Dodeca», 2000 P.35-37

11. Rudenko M.Yu., Alexeyev V.B., Matsyuk S.A. Biophysical Phenomena in Blood Circulation in Indirect Measuring Arterial Pressures & Evaluation of the Relevant Instrumentation // «Medtechnica». 1986 No.5   P.26-35

12. Popetchitelev Ye.P., Korenevsky N.A. Electrophysiological & Photometric Medical Equipment. M., «VysshayaShkola», 2002, P.75-76

13. Cardiology Manual / N.A.Manak, V.M.Alkhimovich, V.N.Gaiduk & et al. Mn.: "Belarus», 2003, 624 pages

14. Voronova O.K. Development of Models & Algorithms of Automated Transport Function of the Cardiovascular System. Doctorate Thesis. Prepared by Mrs.O.K.Voronova, Ph.D., Voronezh, VGTU, 1995, 155 pages

15. Patent No.94031904 (RF). Method of Determination of the Functional Status of the Left Sections of the Heart & their Associated Large Blood Vessels. Authors: Poyedintsev G.M., Voronova O.K.

16. Karpman V.L. Phase Analysis of the Heart Function. M., «Medicina», 1965. 328 pages

17.Fatenkov V.N. Biomechanics of the Heart, Phase Structure of Heart Cycle. Samara, The Samara State Medical University, 1998, 16 pages

18.Kramarenko A.V. RED as Universal Indication of an Affection of the Cardiovascular System [electronic resource]: Access: http://www.dx-telemedicine.com/rus/publications/hubble/hubble Ol.htm. Screen title

19.Bogdanova N.V. Development of Algorithms and Application Software for Cardiologist-Reseaercher's PC-Workstation. Thesis prepared by Mrs.Bogdanova N.V., Ph.D, the Ryazan State Radio-Engineering Academy, presented 27.06.2001

20.Ivanov G.G. High-Resolution Electrocardiography. M., «Triada-X», 1999, 280 pages

21.Gaiton A. The Minute Heart Volume & its Regulation. M., «Medicina», 1969, 472 pages

22.Savitsky N.N. Biophysical Principles of Blood Circulation & Clinical Methods of Researches in Hemodynamics. L., "Medicina», 1974, 312 pages

23.Reference Book. MOTOROLA Sensors

24.Rutkovsky J. Integral Operational Amplifiers. M., «Mir», 1978, P. 180

25.Tchizhevsky A.L. Structural Analysis of Moving Blood. M., Academy of Sciences of the USSR, 1959, P.3

26.Tchizhevsky A.L. Electrical & Magnetic Properties of Erythro-cytes. Kiev, "Naukova Dumka», 1973

27.Regirer S.A. Some Issues of Blood Circulation Hydrodynamics. In a Miscellany of Translations: Blood Circulation Hydrodynamics. M., "Mir», 1971, P.242-258

28.Folkov B., Nil E., Blood Circulation. M., «Medicina», 1976, P.19

29.Cardiology Manual, under Editorship by Ye.I.Tchazov. M., «Medicina», 1982 Vol. 1 P.195

30. Johnsov P., Peripheral Blood Circulation, M., «Medicina», 1982 P.136

31.Remizov A.N. Course in Physics, Electronics & Cybernetics. M., «Vysshaya Shkola», 1982 P.96

32.Rosen P. The Principle of Optimization in Biology. M., "Mir», 1969 P.71

33.Altshul A.D. Losses due to Friction in Piping. M., «Stroyizdat», 1963 P.192

34.Whitmore R.L. Rheology of the Circulation. Oxford: Pergamon Press, 1968

35.Bloch E.H. Amer.J.Anat. 1962.110. No.2. 125

36.Levtov V.A., Regirer S.A., Shadrina N.Kh. Blood Rheology. M., „Medicina», 1982. P.188

37.Shlichting G. Theory of Boundary Layer. M., IL, 1956, P.57

38.Feinmann R., Leuton R., Sands M. Feinmann's Lectures in Physics. M., „Mir», 1977 Vol.1 P.70

39.Turbulency. Principles & Application. Under the editorship of W.Frost, T.Moulden. M.,"Mir», 1960 P.6

40.Bargessyan M.G. About Friction Coefficient under Unsteady-State Movement Conditions in Pipes». Academy of Sc. of the Armenian SSR. Technical Series, M., 1971, Vol.24 No.6

41.Coppel T.A., Liiv U.R. Experimental Investigations of Fluid Movement Initiation in Piping. Published by the Academy of Sciences of the USSR, MZhG, 1977 No.6 P.69

42.Dorodnitsyn A.A. Mathematics & Descriptive Sciences. In a Miscellany of Papers under the Title: Number & Idea. Public.No.5, edited by N.N.Moisseyev. M., "Znaniye", 1982 P.6-15

43.Kisselev P.G. Hydraulics. The Fundamental Principles of Fluid Mechanics. M., "Energiya», 1980, P.295

44.Wiggers C, Blood Circulation Dynamics. M., "IL», 1957, P.78

45.Rashmer R., Dynamics of the Cardiovascular System. M., «Medicina», 1981

46.Seleznyeva S.A., Vashetina S.M., Mazurkevitch G.S., A Comprehensive Evaluation of the Blood Circulation in Experimental Pathology». M., «Medicina», 1976

47.Emmerich J. u.a. Uber den Einfluss blutiger Untersuchungs-methoden auf das Herzminutenvolumen. Zeitschrift „Kreislauf», 1958, Bd.47 S.236

48.Fahraeus R.Physical Rev.1929.9. 231

49.Lipowsky H.H., Kovalcheck S., Zweifach B.W. Circulat.Res. 1978.43 No.5.738

50.Zaretsky V.V., Bobkov V.V., Olbinskaya L.I. Clinical Echocardiography. Atlas. M., „Medicina», 1979

51.Mukharlyamov N.M., Belenkov Yu.N. Ultrasound Diagnostics in Cardiology. M., „Medicina», 1981

52.Karpman V.L., Sinyakov V.S. Three-Dimensional Dynamics of the Left Ventricle and Phase Structure of the Heart Cycle // Physiol. Journal. USSR. 1965

53.Kubicek W.G., Petterson R.P., Witson D.A. et al. Development and Evaluation of an Impedance Cardiac Output System. Aerospace Med. 1966.V.37 No.12 P.1208-1212

54.Kubicek W.G., Petterson R.P., Witson D.A. Impedance Cardiography as a Noninvasive Method of Monitoring Cardiac Function and Other Parameters of the Cardiovascular System. Ann.N.V.Acad. Sci.1970 V.170 P.724-732

55.Tischenko M.I., Smirnov A.D., Danilov L.N. et al. Characteristics and Clinical Application of Integral Rheography - a New Method in Measuring Stroke Volume. "Cardiologiya», 1973

56.Lightfoot E. Phenomena of Transfer in Live Systems. M., «Mir», 1977

57.Bronshtein I.N., Semendyaev K.A. Reference Book in Mathematics. M., «OGIZ", 1948 P.309-328

58.Prischepa M.I. Features of the National Assurance of the Measurement Units in Cardilogical Diagnostics Laboratory [electronic resource]: Laboratory medicine. 2003 No.6 Access: http://www.ramld.ru/articles/article.php?id=37 Screen title

59.Prokopov A.V. Algorithm of Justification of Equation of Measurement & Evaluation of a Methodical Error (Imprecision) in Indirect Measuring [electronic resource]: Access: http://mscsmq.vniim.ru/files/2004/rus/prokopov-ru.pdfScreen Title

60.Rudenko M.Yu. Development of Interferential Method in Measuring Arterial Blood Pressures & Instrumentation Based thereon. Thesis. Summary prepared by Mr.Rudenko M.Yu., Ph.D.. M., MNIIIMT, 1989, P.20

61.Havaa Luvson, Traditional and Current Aspects of the Oriental Reflexotherapy. Edition 2, revised. M., "Nauka», 1990, 576 pages

62.Phase Analysis of the Heart Cycle & Diagnostics Based on CARDIOCODE Application. M.Yu.Rudenko et al. Rostov-on-Don, Published by the Rostov State University, 2005, 56 pages

63.Leninger A. Fundamentals of Biochemistry, 3 Volumes. Translation from English into Russian. M., «Mir», 1985, 1056 pages

64.Kapelko V.I. Disturbance in Energy Generation in Cardiac Muscle Cells: Causes and Consequences // Soros' Education Journal. Biology. 2000. Vol.6 No.5 Pages 14-20 Access: www.pereplet.ru/obrazovanie/stsoros/992.html

65.Kapelko V.I. Creatine Phosphokinase Way of Energy Transport in Sarcous Cells // Soros' Education Journal. Biology. 2000 Vol.6 No. 11 Pages 8-12

66.Rubtsov A.M. The Role of Sarcoplasmatic Reticulum in Regulation of Contraction Activity of Muscles // Soros' Education Journal. Biology. 2000. 39. Pages 17-24 Access: www.pereplet.ru/obrazovanie/stsoros/1066.html

67.Kapelko V.I. Blood Circulation Regulation // Soros' Education Journal. Biology. 1999. No.7 Pages 79-84 Access: http://www.pereplet.ru/cgi/soros/readdb.cgi?f=ST750

68.Garkavi L.Kh. Activation Therapy. Rostov-on-Don. Published by the Rostov State University, 2006, 256 pages

69.Yelisseyev Yu.Yu. Coronary Heart Disease (CHD. Stenocardia. Myocardial Infarction. Pathogenesis, Clinical Picture, Diagnostics & Therapy). M., "Kron Press", 2000, 170 pages

70.Guriya G.T. et al. Mathematical Model of Activation of Blood Coagulation & Thrombus Growth under Blood Circulation Conditions. A Miscellany of Papers under the title "Computer Simulations & Progress in Health Service», M. 300 pages 'Nauka», 2001,

Contents